Sigurd Bergmann, Dieter Gerten (Eds.)

Religion and Dangerous Environmental Change

Studies in Religion and the Environment
Studien zur Religion und Umwelt

published on behalf of the
European Forum for the Study
of Religion and the Environment
by

Sigurd Bergmann

Volume 2

LIT

Sigurd Bergmann, Dieter Gerten (Eds.)

Religion and Dangerous Environmental Change

Transdisciplinary Perspectives on the Ethics of Climate and Sustainability

LIT

Cover Picture: Hardy Brix, *Arktisk Landskab*, 2008 (above)
Hardy Brix, *Arktisk Landskab 2*, 2006 (below)
Hardy Brix, *Arktisk Landskab 4*, 2006 (back cover)

Gedruckt auf alterungsbeständigem Werkdruckpapier entsprechend
ANSI Z3948 DIN ISO 9706

Bibliographic information published by the Deutsche Nationalbibliothek
The Deutsche Nationalbibliothek lists this publication in the Deutsche Nationalbibliografie; detailed bibliographic data are available in the Internet at http://dnb.d-nb.de.

ISBN 978-3-643-10093-1

A catalogue record for this book is available from the British Library

Fresnostr. 2 D-48159 Münster
Tel. +49 (0) 2 51-620 320 Fax +49 (0) 2 51-922 60 99
e-Mail: lit@lit-verlag.de http://www.lit-verlag.de

Distribution:
In Germany: LIT Verlag Fresnostr. 2, D-48159 Münster
Tel. +49 (0) 2 51-620 32 22, Fax +49 (0) 2 51-922 60 99, e-mail: vertrieb@lit-verlag.de
In Austria: Medienlogistik Pichler-ÖBZ, e-mail: mlo@medien-logistik.at
In Switzerland: B + M Buch- und Medienvertrieb, e-mail: order@buch-medien.ch

In the UK: Global Book Marketing, e-mail: mo@centralbooks.com

In North America by:

Transaction Publishers
New Brunswick (U.S.A.) and London (U.K.)

Transaction Publishers
Rutgers University
35 Berrue Circle
Piscataway, NJ 08854

Phone: +1 (732) 445 - 2280
Fax: + 1 (732) 445 - 3138
for orders (U. S. only):
toll free (888) 999 - 6778
e-mail: orders@transactionpub.com

Contents

Religion in Climate and Environmental Change

Towards a Symphony of Voices, Memories and Visions in a New Polycentric Field

Sigurd Bergmann and Dieter Gerten

Our common past – our common future

In 1857 John Stuart Mill asked the questions: "To what final goal will technical progress lead humanity? In what state of being will it exist at the end of this process?"

This quote from Mill, the influential Scottish economist and philosopher (1806-1873), is given by the German land artist Hermann Prigann (1942-2008) in his reflections about ecological aesthetics.[1] Although Mill himself was not at all interested in questioning technical progress with regard to his overall theory about the free market's capacity to allocate resources for the benefit of all, Prigann uses Mill as a starting point to reconstruct his own artistic work. From a hermeneutic point of view one might shake one's head but artistically speaking one should not raise one's hackles: Mill was among the architects of modern economy and the growth paradigm, the underlying ethics and practices of which have eventually catalysed anthropogenic climate change and other life-threatening environmental changes on a global scale. Prigann, in contrast, belongs to the increasing group of creative perceivers, thinkers and actors who experiment with places, artefacts and paths that might lead humanity and the planet Earth through the sufferings of environmental change. In a couple of highly provocative, widely acknowledged projects the artist has reconstructed urban and industrial landscapes and transfigured them into exciting places of remembrance, contemplation and ritualisation.

Why do we refer to Mill and Prigann in the introduction to this book about "religion in climate and environmental change"? Because we think that this book's theme is also at the heart of Prigann's work, if we replace the term "religion" with "arts". Let us therefore turn to the arts – and to Prigann's Land Art in particular – before diving into the wide sea of religion(s).

In his long-term project "Terra Nova", Prigann designed and built his "Himmelstreppe" (Stairs to Heaven) at the site of a dump at an abandoned coal mine near the

1 Hermann Prigann, Himmelstreppe (stairs to heaven), on top of the dump environment of the colliery area Rheinelbe, now the "Skulpturenwald Rhein-Elbe", 1999, © Herman Prigann / BONO, Oslo 2010, photo: © Thomas Robbin

German Ruhrgebiet. Fragments and debris from the built places are interwoven with new and organic materials. This way, the archaeological past does not fade away but receives new meaning and function. The lonesome artefacts point both backwards and forwards in time, and as they remain in place the historical wounds on the landscape begin to heal. Aesthetically, Prigann alludes to Mesoamerican ritualistic places as known from, for example, Mayan sacred geography.[2] The religious transfer from one context to another invites us to ritualise the former mine anew. What kind of social actions could and should take place here and now? What kind of awareness with regard to human skills of destruction and construction might emerge? The restoration, or better, transfiguration of the mine into a new open eco-spiritual place is furthermore embedded into a larger master plan for a landscape park, where extensive areas of former industrial use begin to be turned into a landscape whose future appearance is far from clear and predictable.[3] Land Art thus offers a specific quality of change and uncertainty, which might teach us something about the tremble und tremor in social and natural disasters catalysed by dangerous environmental change. One of the challenges in the context of climate change is to differentiate between those uncertainties that potentially threaten life and those that can encourage us to increase our dignity for life.

The artist completes his stairs to heaven with the Great Chair ("Der große

Stuhl"), which references an archaic throne but is built from fragments of granite and stone from the old mining company's central building. The "Spiralberg" establishes with the Stairs to Heaven a mythical landmark which embeds the visitor and provides him/her with a means of orientation in the larger surrounding, and also invites him/her to reshape the significance of the place by physically moving around and mentally creating new inner images. The "sins" of the past – represented here by the former mine whose products contributed to the CO_2 emissions that eventually force global climate change – are in this way slowly transformed into landscapes where change can take place in a peacefully reconciled, non-threatening way.

2 Hermann Prigann, Waterlevel, transformation of a former water-pumping station (near Marl, Germany) into a landscape artwork, 2001, © Herman Prigann / BONO, Oslo 2010, photo: © Herman Prigann

Prigann's art furthermore reminds us of the fact that anthropogenic environmental change does not only encompass climate change but also global and regional land use changes (that is, the widespread conversion of natural ecosystems to arable land, pastures, settlements, etc.); related disturbances of terrestrial and aquatic ecosystems with threats to faunal and floral biodiversity; contaminations of water and soil; and many others.[4] He might in this vein offer us pause for thought in his restoration of an abandoned water work, where the natural flow of the river has become visible and tactile again (cf. image 2). The hill thus "carries" the ruin, and the landscape embraces our common past and opens our common future. This may remind the passer-by of the fact that without intact aquatic and terrestrial ecosystems and without a fair and sustainable use of freshwater, conflicts between human societies may become more and more insecure, violent, conflict-laden and painful.[5] A global water ethos is, as Dieter Gerten shows in his chapter of this book, as essential for establishing patterns of creative adaptation to climate and environmental change as it is to turn the former CO_2-intensive coal mines into flourishing gardens of redemption.

Another crucial challenge, visualised by the artist's transformed site, is the need to make oneself at home again in a world that is permanently being destroyed by humans themselves. What in general will home (German "Heimat") mean, and how will processes of "earthing and making-oneself-at-home" develop in the context of accelerating and expanding environmental change and the related changes in lifestyles, belief patterns and modes of surviving? How can religion itself contribute to the earthing and "Beheimatung" in this world?[6] Or, to put it in Anne Primavesi's words from the title of her chapter, what does it mean to be alive with Gaia in a relational world?

In Prigann's sense the present book implicitly asks the following questions: Will our common global future really be determined by the collective faults of our common past – as projected by climate science and visualised in many kinds of graphs that show first slowly then rapidly increasing curves of change in global temperature and resulting impacts upon human societies and ecosystems? Or can we create qualitatively different futures by actively, wisely and creatively shaping the atmosphere, the landscapes and the global water system in sustainable manners – like Prigann does in his Land Art? Hence, will we be able to create a world in which mitigation curves straighten out, or will the need for adaptation increase for generations because we decide to continue with business as usual? If we are heading for alternatives, how should we renegotiate the connections and interplays of science, politics, religion and culture for a sustainable future – not in a narrow short-term but in a long-term, deep sense? Can economy and technology alone "save" us as has been believed for centuries, or do we need to revise even these foundations of modern societies so as to prepare for the necessary breakthroughs?[7] Where can the spiritual, cultural and artistic resources for this be found, and how can they be mobilised in a quantity and quality never needed before?

The authors of this book are especially interested in the religious dimension of the cultural mobilisation that is ultimately needed for a global sustainability transition. Specifically, they address the following questions:

- What is "religion" in the context of accelerating and expanding global environmental change? What concepts of religion are fruitful and how do they connect to environmental change?
- How does religion work in critical (catastrophic) processes of environmental change? More specifically, what can religion contribute to the adaption to such

change, and how may it hamper the mitigation of change and sustainable development in general?
- Further, in what ways does climate change alter religious institutions and images of the Sacred?
- Which cultural predicaments contribute(d) presently and historically to global climate/environmental change?

The reader should not expect quintessential answers to all of these questions, as our intention in the first place is to offer solid formulations and revised approaches to the field of religion in environmental change so as to prepare for a longer-term and deep cultivation of this research field.

Creative Adaptation with a Human Face and Soul – Why Religion Matters

Given the massive, worldwide problems that are to be expected under climate and environmental change and the fact that a large part of the world's population is not (yet) subject to the technocratic worldview, it is imperative to complement the existing technological and economically oriented problem solutions with alternative perspectives – narratives that integrate the entanglement of humans and their environment.

Not only religion itself but also the understanding of science in general is contested in the context of anthropogenic environmental and climate change. The dialogue and cooperation with religious studies and cultural studies can be highly fruitful for increasing the self-critical transformation of natural science into a truly transdisciplinary environmental science. Although climate science in itself has written a successful story of creative interdisciplinary cooperation, it still represents an academic practice "without human faces", i.e. with insufficient integration of the humanities. The IPCC process offers a similarly significant success on the political level, but it must exactly for this reason, as many argue now, be taken further to a genuine cooperation of all faculties in the future. The inclusion of the humanities and social sciences as equal partners seems to be crucial for such a necessary paradigmatic change. We regard "religion" as a microcosm for the natural and cultural state of affairs today, where human skills to perceive, believe, think and act as bodily as well as spiritual creatures moves to the forefront of such an envisioned synthetic environmental science. That said, the present book offers a contribution to such a renegotiation of what science is and could be, in general as well as particularly with regard to our common future in global environmental change.

There is a large body of evidence supporting the fact that "religion" indeed can serve as a cultural microcosm within which humans act in environmental/climate change, as follows.

- The majority of the world's inhabitants are practicing religious believers. A socio-cultural analysis of practices, values and worldviews with regard to dangerous environmental change must therefore necessarily also include the religious dimension with its ethical, aesthetical and political ambiguities.
- It is contradictory to pull systemic solutions over large parts of local populations without involving them as central agents of adaptation and mitigation. An analysis of the complex interplay of religion and environment produces valuable insights into the cultural dynamics of global environmental change.
- Images of the Sacred – or "ultimate sacred postulates" (Roy R. Rappaport); that is, values, norms, visions, cosmologies and constructions of meaning – are always at work in human actions, also in secular social processes such as financial capital accumulation or environmental politics. Religion can therefore be studied as a phenomenon in both the generation of environmental changes and in the responses to them. Fabrications of meaning are furthermore strongly involved in the public discourse about climatic change.
- Faith communities have in the last few years developed rapidly increasing and intense activity with regard to the mitigation of anthropogenic climatic change. For example, evangelical Christians in the USA have divided into two camps: one arguing for the urgent moral challenge of climatic change; and the other using the fear of climate change to promote a new pagan religion similar to the medieval fear of hell. Similarly, there are many religious and ethical incentives and actions for a more sustainable use of the Earth's scarce freshwater resources.[8]
- Accordingly, churches and other religious bodies intervene in the public spheres with declarations about the spiritual dimension of a civilisation in crisis.[9] However, the kind of political influence that religious leaders and believers develop has hardly been researched until now.
- At least some religions emphasise the ethical dimension and the geopolitical and ecological injustice, which becomes even more inescapable in ongoing global climate change and its asymmetrical distribution of risks and sufferings between rich and poor. The ethical notion of "ecological justice" has been debated intensively with respect to relations between humans and non-humans, with respect to the relations between the poor and the rich and with respect to the locally bound and the global players. The developing concept of "climate justice" fits particularly well into this discourse.
- Furthermore, "religion" is not only a function in the process of the mobilisation of citizens, but an analytic approach to religion, in Religious Studies and Theology as well as in affiliated disciplines, offers tools for the critical analysis of the deeper driving forces implicit in the social dynamics of *producing* environmental change. The metaphysics and dogma of economic growth and accumulation, the anthropocentric self-understanding of humans as managers of nature rather than as part

of it, the ideological construction of technology as a simple tool for salvational manipulation, the ethical and spiritual driving forces in the environmental movement (parts of which sociologically act like faith communities), the emergence of eco-citizenship and ecological spirituality: All these phenomena have been explored for 30 years in the research field of "Religion and the Environment", and four significant international networks have been established in the USA, Canada and Europe. Scholars in Asia and Africa are also on their way.[10]

The Disposition

The chapters of this book are revised presentations from two workshops with members of the European Forum for the Study of Religion and the Environment (EFSRE), held at the Potsdam Institute for Climate Impact Research (Potsdam, Germany) in December 2007 and at the Norwegian University of Science and Technology (Trondheim, Norway) in October 2008. A third workshop in this series has taken place in Potsdam in January 2010 on the theme of "Religion in Global Environmental and Climate Change: Sufferings, Values, Lifestyles", and its results will be shared in a forthcoming book. From the beginning it was our intention not to stay in the narrow rooms and corridors of familiar and comfortable academic territories but to interconnect different perspectives on the phenomenon of "religion and environment in change". The chapters therefore invite readers to share contributions from such different disciplines as climatology/hydrology, religious studies, theology, cultural studies, philosophy and visual arts. In the following, we provide a brief overview and synthesis of these contributions.

In his introductory essay *Sigurd Bergmann* departs from the hypothesis that climate and environmental change alters religion, and he then meanders through a multifaceted landscape of the interconnections between science and religion. As a golden thread running throughout is the question of how physical perceptions of the environment and its changes are related to images and spiritual fabrications of meaning of nature, and how the perceived, the imagined and the practised are intertwined in what Jakob von Uexküll has interpreted as the circle of meaning and action. Religion works in such a frame as "lived religion in, with and through lived space", that is as a both ideational and practical, as a both ethical and aesthetical dynamics at the heart of human ecologies. Finally, Bergmann marks key aspects for future research in the field of "religion in climate and environmental change".

Owing to its vital importance for all life on Earth, water plays a key role in almost all religions and is the medium of numerous religious rituals and practices. For this reason, hydrologist and geographer *Dieter Gerten* focuses on water when reviewing the manifold ways in which religion shapes humans' interactions with their environment – and vice versa in which ways religion can be shaped by environmental changes. However, as the author demonstrates by means of case studies from different corners of the world, the influence of religion on water use is highly ambivalent: On the one hand, water is indeed often regarded as a sacred element that merits careful use and protection, and religion can help people to cope with natural disasters such as floods. On the other hand, there are many examples that theoretical theological considerations do not necessarily play a role in practice, and that the influence of religious views and practices may even oppose the (Western) view on how fresh water should be used sustainably.

Water, in all its forms and aggregate states, is also at the core of the Danish artist *Hardy Brix'* three painted landscapes included in this book. These originate from the painter's encounter with arctic and subarctic surroundings on the deck of a slow-moving research vessel in an environment that overwhelms our imaginative and reflexive skills. As the artist has visited these waters for many years, the paintings displayed in this book are by no means snapshots but impressions and expressions of an enduring deep encounter with a land- and seascape in continuous change.

Petra Hansson's essay, thereupon, approaches Hardy Brix' painted land- and waterscapes from the angle of literary and eco-criticism. The Swedish educationalist and literary scholar discusses what we can learn from artistic narratives in relation to technological and scientific narratives of global climate change. According to her argumentation, education and education research can provide very important contributions to the public and academic debates about sustainability and climate change, with significant potentials to increase citizens' responsibility in learning about the ongoing environmental changes. Hansson investigates literary critic Louise Rosenblatt's theory of "transactional reading" and applies it to Brix' landscapes. Following Rosenblatt's theory and her ideas of "aesthetic" and "efferent" reading stances, two different interpretations of the paintings are proposed. Even though literary criticism is mainly concerned with responses to, and readings of, written texts, Hansson's investigation manifests the relevance and unique contribution of aesthetic eco-criticism to climate and environmental science.

Changes in landscapes, and especially their vulnerabilities to anthropogenic influences, are also the focus of the chapter by Canadian philosopher *Thomas Heyd*. He points out that landscape is often not conceived as inert but as *sentient*, with a consequent sense of "social responsibility" for landscape. After a brief discussion of the notion of vulnerability with regard to natural phenomena, he discusses different ways in which natural life has been perceived, and how the sense of responsibility toward landscape develops among peoples who are deeply rooted in the landscape that they inhabit. What underlies this sense of responsibility? Heyd makes us also aware of how respect for natural phenomena may lessen environmental vulnerability, and he concludes that it is imperative to develop those ways of conceiving natural phenomena that lead to a deep sense of respect and responsibility for the natural world that surrounds us.

Anne Primavesi uses the term "vulnerability" in a different context: Building on Catherine Keller's notion of Mutually Assured Vulnerability and on Lovelock's assertion that we live in a Gaian world, she pleas for nonviolent ways to counteract global climate change (by avoiding as far as possible injury to others such as the poor) as opposed to the often inherently violent desire to dominate and exploit natural resources and people for one's own profit. Starting from a Christian perspective she develops this further by means of a highly original interpretation of Dietrich Bonhoeffer's thoughts.

For British theologian and ethicist *Michael Northcott,* trees play an essential role in the process of climate change. As we know, anthropogenic climate change involves the displacement of stored subterranean carbon from under the earth's surface into the atmosphere – he focuses on the fact that most of this carbon had been drawn down from the atmosphere by trees through photosynthesis, and that living trees are now threatened not only by climate change but also by replacements for bioenergy production and food and fibre production. In this context, Northcott reminds us of the fact that especially in Islam and Christianity trees symbolise the nature of divine truth, and that the human relationship with trees is a core motif of religious criticism of human waywardness. Recovering a conception of trees as sacred may be an important step in resisting the turn of all forests, subterranean and still living, into mere standing reserves of energy whether as carbon or biofuel.

Timothy B. Leduc's chapter calls for a shift in climate and environmental science toward a more spirited transdisciplinary und multipolar direction, which would require a reconnection of knowledge from the various scientific disciplines, environmental policies and history with the humanities (including the religious sciences). He briefly charts the spread of interdisciplinary environmental research in historical times (focused on Western thinking), and reflects upon the nature of an enhanced environmental science. For this, Leduc builds on his own pilot studies in which he has carefully considered the traditional ecological knowledge of Inuit as an equally valid interpretation of environmental changes.

Italian-German cultural scientist *Davide Brocchi* addresses from an even wider perspective the cultural dimensions of present-time (un)sustainability. Starting from Jared Diamond's theory that societies might collapse if they do not manage sustainably their environmental foundations, he states that (modern) societies – and particularly the neoliberal way of thinking – can be characterised as being self-referential as they tend to externalise disorder, "useless" things, contradictions and alternatives from the social system. Thence he develops principles for diverse cultures of sustainability, which shall include, among others, the recognition of the interconnectedness of all things, the awareness of limits to growth, and the encouragement to redefine personal freedom as a freedom of responsibility: for example, the freedom to disobey unsustainable lifestyles in the spirit of a positive understanding of humanity. Finally, he develops elements of a socio-cultural evolution towards sustainability – which he regards as a systemic and structural alternative to neoliberal globalisation – and also reflects on the role that Christianity may (or may not) play in this process.

Public discourses indicate that researchers, faith communities, economists, lay-people, climate experts, policymakers, artists and educators agree that climate change involves serious moral challenges to human societies. Nevertheless, with some exceptions ethicists have not yet paid enough attention to the moral dimensions of anthropogenic climate change. Given this evidence, the purpose of the chapter by Swedish ethicist and educationalist *David Olsson Kronlid* is to identify and clarify the ethical considerations in climate change science as reflected in the Fourth Assessment Reports of the Intergovernmental Panel on Climate Change (IPCC). By applying and developing further Martha Nussbaum's elaborated "capabilities approach" he highlights that dangerous climate change threatens not only valuable resources and people's welfare but also the qualitative dimensions of human wellbeing, i.e. the abilities that people need for leading a valuable life – such as appreciation of beauty, self-determination, transcendence, and compassion for non-human species. Kronlid finds that these dimensions are hardly if at all addressed in the IPCC reports, and he suggests how an application of the capabilities approach might help to systematically integrate the ethical dimensions in climate change science.

In 48 points, or better "steps", German philosopher and environmental ethicist *Konrad Ott* continues Kronlid's struggle with the ethical implications of climate change by elaborating basic foundations on existing and necessary policies for climate change. The short sections are consistently formulated as "ethical claims" and the reader should approach these slowly and with concentration, so that the subsequent steps are converted into one single walk and path. At the core of the author's argument lies the climate ethics concept of "Contraction and Convergence". This argument provokes a constructive debate, and it presumes to support a concept that has been regarded as being utopian a decade ago but now has entered the political stage. What might it contribute to international climate policy in a non-distant future?

"What are we Waiting For?" and "How do we resist the temptations of climate despair?" asks finally British philosopher of religion *Stefan Skrimshire*, at the same time orienting us towards the apocalyptic end of space and time. According to him, climate change discourse – and especially the notion of "tipping points" in the Earth system – draws deeply on metaphors of irretrievable and in some way inevitable loss: of humanity's time running out; of the earth catching a morbid fever; of Gaia taking her revenge. Skrimshire reflects that stories of the end of life can have as much power and influence on the moral imagination as the facts and model projections of climate science. There has thus never been a more important time to appreciate the role of religious narrative in communicating crises of the future, and in particular the contemporary fascination with the possibility of human extinction through climate change.

However diverse the present book chapters are – they are all guided by a quest for alternative pathways that may lead humanity through the era of anthropogenic climate change and other environmental crises, and also by the search for a deeper understanding of the causes and consequences of, and possible responses to, these crises. They highlight that there are good reasons to not just *add* religious and ethical perspectives to the natural and socioeconomic sciences (which presently dominate the environmental/climate change discourse) but to fundamentally *revise the significance* of these perspectives for to find both ideational and practical solutions to the increasingly dangerous environmental changes. Of course, religious approaches are not proposed to be the one and only solution to these global changes and crises – the authors acknowledge ambivalences and carefully consider the both positive and negative aspects of religion in environmental change. Eventually, the present analyses demand for a rigorous and comparative synthesis of the natural sciences' perspectives with the manifold theological, ethical and philosophical perspectives. Without such a plurality of views on climate change, nature and our common future[11] the "one-dimensional" humanity (Herbert Marcuse) can hardly find its way out of the present self-referential systems.[12] As editors we would be glad and proud if this book contributed in some small way to widening the panorama of challenges and skills that are needed for a turn to multi- and transdisciplinary environmental science and ultimately to an overall transition towards sustainability.

Notes

1 Hermann Prigann, Wien, 30 August 1986, quoted in: Strelow, 32. Cf. Prigann's own comments about his work: "Destroyed landscapes are activity fields available for creative remodelling. It is necessary to create the preconditions for resource yielding landscapes where energy raw materials were extracted. A failure to use human energy and creativity in work for these goals would throw doubt over the long-term survival of communities. Without an aesthetic and visionary starting point for the reshaping of the destroyed areas, there can be no positive identification with the landscape. The aesthetic reshaping of the landscape is always orientated around the Ecological relationships, in this sense we are talking about ecological aesthetic." Hermann Prigann, <http://www.terranova.ws/tnvision.htm>, accessed 26 October 2009.

2 Cf. Bergmann, chapter 3.

3 The "Emscher Landschaftspark" forms, with the involvement of twenty cities, a part of the broader transformation of the central German industrial region "Ruhrgebiet" into a so called post-industrial cultural landscape. Cf. <http://www.rvr-online.de/landschaft/Emscher_Landschaftspark/emscherlandschaftspark.php>.

4 For a brief overview of the global environmental problems and the need for a sustainabil-

ity science and sustainability transformation of societies, see e.g. Kates.

5 See e.g. Delli Priscoli and Wolf.

6 Cf. Bergmann.

7 In a memorandum of Nobel Laureates it is stated that "climate protection ambitions appear to be on a collision course with the predominant growth paradigm that disconnects human welfare from the capacity of the Planet to sustain growth" (Potsdam Memorandum 2007, <http://www.nobel-cause.de/Potsdam%20Memorandum_eng.pdf>; see also a further memorandum signed by 59 Nobel Laureates, <http://www.nobelcause.org/>.

8 See especially Chamberlain.

9 Some representative examples of this increasing activity in the traditional religious and also new spiritual spheres are:

– The Interfaith Climate Summit in Uppsala 2008 gathered religious leaders from many traditions all over the planet; it's manifesto is available at: <http://www.svenskakyrkan.se/default.aspx?di=173312>

– Muslims have created a global network of exchange on environmental issues, the Islamic Foundation for Ecology and Environmental Sciences (IFEES), cf. <http://ifees.org.uk/>

– International ecumenical institutions such as the World Council of Churches and the Lutheran World Federation have contributed to the discourse and mobilised believers, mainly in the poor world regions. Cf. <http://www.wcc-coe.org/wcc/what/jpc/earthdocs.html> and Bloomquist.

10 See <http://www.hf.ntnu.no/relnateur/index.php?lenke=links.php>.

11 Cf. Leduc.

12 See Gerten.

References

Bergmann, Sigurd (2010), *Raum und Geist: Zur Erdung und Beheimatung der Religion – eine theologische Ästh/Ethik des Raums*, (Research in Contemporary Religion 7), Göttingen: Vandenhoeck & Ruprecht.

Bloomquist, Karen L. (ed.) (2009), *God, Creation and Climate Change: Spiritual and Ethical Perspectives*, (LWF Studies 2/2009), Geneva: The Lutheran World Federation.

Chamberlain, Gary L. (2007), *Troubled Waters: Religion, Ethics, and the Global Water Crisis*, Lanham: Rowman & Littlefield.

Delli Priscoli, Jerome and Wolf, Aaron T. (eds.) (2009), *Managing and Transforming Water Conflicts*, Cambridge: Cambridge University Press.

Gerten, Dieter, "The noosphere in Earth system analysis", presentation at the International Climate Congress "Climate Change: Global Risks, Challenges & Decisions", 10-12 March 2009, Copenhagen.

Kates, Robert W. (2001), "Queries on the human use of the Earth", *Annual Review of Energy and the Environment,* Vol. 26, 1-26.

Leduc, Timothy B. (2007), "Sila dialogues on climate change: Inuit wisdom for a cross-cultural interdisciplinarity", *Climatic Change*, Vol. 85, 237-250.

Strelow, Heike (2004), *Ökologische Ästhetik: Theorie und Praxis künstlerischer Umweltgestaltung,* Basel: Birkhäuser.

Dangerous Environmental Change and Religion

How Climate Discourse Changes the Perception of our Environment, the Spiritual Fabrication of its Meaning and the Interaction of Science and Religion

Sigurd Bergmann

Climate change transforms the interaction of science and religion

"Climate is a kind of religion", Nigel Lawson, former British finance minister for Margaret Thatcher, declared in a lecture,[1] where he compared eco-fundamentalism to Islamist terrorism and bemoans the decline of reason. Successful science writer Michael Crichton, who has fuelled the criticisms of climate science in his novel "State of Fear" says it even more clearly:

> One of the most powerful religions in the Western World is environmentalism. Environmentalism seems to be the religion of choice for urban atheists.[2]

The notion of "religion" is often used for to discriminate against insights and conclusions drawn from the new consciousness of climate change. Radical conservatives portray G.H. Brundtland, with her mandate from the UN, as a "bishop in climate religion", and evangelical Christians in the USA have been divided into two camps, one arguing for the urgent moral challenge of climate change and demanding "an urgent call for action",[3] while the other is using the fear of climate change to promote a new pagan religion similar to the medieval fear of hell. Also sceptical scientists act as naysayers with regard to the mixing of science and religion.

The use of the term "religion" is interesting in these kinds of polemics and it seems to be fuelled by neoliberal as well as hyperanthropocentric negative attitudes. It can, however, hardly claim any authenticity in accordance with the self-understanding of faith congregations and believers themselves. Their declarations are clear enough about the spiritual dimension of a serious civilisation crisis.[4] Viewed from a Religious studies perspective, both the polemically pro and contra approaches offer strong evidence for the potential of the climate change discourse to evoke and accelerate religious processes. While religion in the first type is imaged as the dangerous opposite of reason, which one needs to get rid of, the second type develops it into a necessary condition for a wisdom inquiry with the potential to enlighten both science and politics.

The intention of this chapter is to offer preliminary reflections about the study of such an interplay of science and religion within the framework of climate change discourse – and in a broader framework the discourse about so called "dangerous environmental change" – and to encourage the further construction of an investigation of implicit and explicit driving forces in this discourse. "Religion" could be studied – independent from the theoretician's position in substantialist or functionalist academic camps – both in the *production* of environmental impacts and in the *solution* of global warming, climate and water problems, such as changing life styles and modes of producing and consuming. Both should, of course, be emphasized, my focus here, however, is on *the construction of the ongoing discourse* and the question how an analysis from Religious studies and Theology could clarify and strengthen its dynamics.

In the following I will meander through five sceneries of consideration.

- Why is the human dimension in general, and the spiritual dimension in particular, crucial for the understanding and solution of environmental problems?
- How does environmental change change religion?
- What can a "lived-space"-approach to religion and the environment contribute to research?
- How can we explore environmental change through the lens of ritual meaning fabrication and the remembrance of suffering?
- How can a research agenda in the field emerge?

The Significance of the Human Dimension and Spiritual Driving Forces in Global Change

The problems emerging from anthropogenic environmental change on global and local scales are located in a broader history of human agency with nature. Modernity and its many "blessings" have led to a historically unique human disembeddedness from natural environments. Science and technology developments in combination with a deregulated global finance market have thrown late modernity into a state of affairs which would need some kind of cultural revolution to find its way out of. The roots of this global ecological crisis, which has been debated now for over thirty years, and which had already been noted by observant scientists in the early 19th century, can be described as sociocultural as well as spiritual.

One of many signs of an awakening is found in Max Schulze-Sölde's "Age of Technology", which displays, strikingly and critically, the polluted life world and poor urban environment of industrialised city space:

1 Max Schulze-Sölde, Zeit der Technik, 1925, oil on canvas, 123 x 94 cm, Kunsthalle Recklinghausen

We could characterise Western civilisation in comparison with others by locating the concept of *nature* at the very centre of its self-understanding. Citizens have identified themselves as parts of nature internally and externally, and they have also acted as rulers and/or guardians of it. The notions of "physis" and "natura" have made it possible to develop connections between the subjective and sociocultural spheres on the one hand and the surrounding space of life on the other. Especially with regard to the moral understanding of reality, the doctrine of "natural law" represents a basic code in European culture by determining an inner connection between moral order and natural order that has been reflected in fundamental theories of ethics and law which still form a strong civic foundation for the process of European integration.

Values, practices and worldviews of citizens worldwide with regard to nature have changed in the last decades. The shift from material to subjectively, culturally and spiritually defined values represents a complex diversity of social forms, practises and ideologies, that extends from animal rights commitments, experiential lifestyles and spiritualities, and health practices to ritual composting, green political parties and ecological globalisation critiques. Climate change consciousness accelerates this process even more at present. Religious ideas and patterns of behaviour are reciprocally connected with images of nature and environment. Religiously grounded values, life-interpretations and images of the future are important for lifestyles, and for the realisation of individuals' and groups' alternative sustainable use of nature.[5] Environmental problems penetrate many sectors of society, in many different patterns of production and consumption, in what von Weizsäcker rightly declared as a millennium of the environment after the past millennium of the economy.[6] Political science has analysed the emergence of a kind of global "ecological citizenship".[7] The process of a transition into a environment-based global culture seems to be beginning.

A widely accepted consensus in environmental ethics, which also is true for climate ethics, is that socially produced problems in the interaction of human beings and nature cannot be solved only through technical or social engineering. Environmental science in general and climate research in particular needs the competence to investigate the deeper forces and processes of empowerment for a long-term sustainable lifestyle for individuals and populations.[8] Environmental problems are in the first instance problems of human and cultural constructions, and the technical and material forms of a non-sustainable eco-management are interwoven with and expressions of human values, attitudes, ultimate concerns and power holding visions as they can be analysed by humanities and especially Theology and Religious studies in a complex way.

Sociological studies show for example how "metaphysical values" contribute to making human views of nature and actions in it more consistent, enduring, and enlarged. New religious movements give a very high value to natural spirituality, and the new forms of religiosity express a deep social and political alienation in a context where the structure of plausibility in the established social institutions and the public citizenship are radically threatened. The sacred dimension for an individual construction of meaning is set free in this process, and it is supported especially through the natural dimension.

Also established religious traditions and institutions have responded to the discourse on the environment and they continue with regard to global climate change. Motorised mobility criticisms have been developed at the forefront by Ecumenical Christian bodies in Europe.[9] The criticism of consumerism as an idolatry in faith and practice has been highlighted in combination with a long-term commitment to the oceanic life world by the Eastern church leader Patriarch Bartholomew, and several church leaders have recently published well informed public declarations. The UNEP has for many years included religious denominations in their work.

One should however not simply regard social movements and faith congregations as pure carriers of social transformation, which in its content builds on scientific records, as is the case in the Wuppertal Institute's report about "Zukunftsfähiges Deutschland".[10] Also the recently published working programme for climate research of the EU is characterised by its exclusion of human (cultural, ethical and spiritual) dimensions and is simply interested in monitoring and technical and socioeconomic engineering of solution policies.[11] Instead of such a blinkered reductionist approach, unfortunately still common, one should regard the social dynamics of environmentalism and the social and spiritual commitment to climate change as a creative force in itself which can affect the design of research as well as the further transformation of politics and the economy. The human religious dimension of this process is however not acknowledged in its depth. As a preliminary hypothesis for our purposes, I would argue that we can identify a deep foundational desire for an alternative way of living beyond the present state of captiveness in social economic and technical bonds, and that it is this desire that produces a creative spiritual force for social transformation. To put it into an early Christian language, climate change makes it necessary to mine our vision of the Spirit who acts for "the life of the coming world", as it is formulated in the ecumenical creed 381 (NC). It also calls for the love to the poor as the highest form of love to God, oneself and one's neighbour as the early Christians have coined it. As Michael Northcott has shown, climatic change deepens the injustice between rich and poor, and without the strong participation of local communities climate politics is doomed to fail.[12]

In the context of climate change, the challenge is to mobilize the spiritual energies hiding in this desire and to interconnect them with a safe scientific state of information and a social arena where experiments towards alternatives could take place. Deep ecologist Arne Næss coined in 1974 the term "ecosophy" for this kind of an emulation, where ecology, as a scientific activity, and sophiology, as a kind of rational belief system, interact without being dissolved into each other.[13] Philosopher

Nicholas Maxwell has expressed a similar approach in his criticisms of the philosophy of knowledge, which does not take into account the living beings self interests and interplay, and which only seeks simple rationalisations of selected phenomena. Instead, he argues, we should replace knowledge with wisdom inquiry. A philosophy of wisdom is for him a truth seeking discourse that strives to negotiate the criteria of why we should act or abstain.[14]

Approaching climate change from the perspective of Religious studies, the challenge is to explore the desire and spiritual driving forces implicit and sometimes explicit in environmental commitment and to widen the present discourse within the framework of wisdom based reflection about why we should act or abstain. While the economistic worldview today only allows an argument, such as Stern's,[15] that we will incur greater costs in years to come if we do not respond today, a wisdom based discourse would explore reasons why local and translocal perceptions of environments are including and producing knowledge and practical skills to survive in a sustainable surrounding where life is treated with dignity.

A reflection of the dimensions of values, life styles, cultural practices, worldviews, and life-interpretations becomes therefore necessary as a key to the solution of the problems that arise in complex self-dynamic patterns of technical, social and economic development.

A central code in all religion is about its universal and translocal validity. The visible hides something invisible powerful, such as in animism, totemism or so called aboriginal religion. Religious traditions in Asia operate with a complex structure of divinities or energies that shape and rule over cosmic processes, while so called monotheist religions in the West regard the one God as the Creator, Sustainer and Liberator of the world, which is perceived as a creation filled with signs, traces and synergies by him/her. Nature represents in this sense the source of revelation, and the Earth and the life of its inhabitants is regarded as a gift which humans never could manage by their own merits. The Hebrew bible's wisdom tradition expresses this clearly and states: The Earth is the Lord's (Exodus 9:29, Psalm 24:1). Similar expressions are found in many religions, and even Buddhism develops at present a practice of ecological awareness that is attractive not only in the East.[16]

A traditional attitude to threatening changes in the natural environment is the belief that God (or the spirits) are reacting to the morally sinful behaviours of human beings. Droughts causing starvation in late antiquity are for example interpreted by the Cappadocian church father Gregory of Nazianzus as a reaction by the Trinitarian God who, as he puts it, "preaches through creation".[17] Augustine in the West has imaged nature as a "liber naturae", beside that of the book of the bible, "liber scriptura", and for him – and the whole medieval tradition until the renaissance – nature and the bible represented two different but equal books in which one could decipher the law and love of the Creator for his creatures. Natural science, for example for Paracelsus, was a semiotical hermeneutics of the Creator's signs in creation.

Science has itself been developed in the medieval scholastic tradition as an interpretation of this invisible uncreated God in the appearance of nature and its laws. While antiquity has used rational explanation as a tool for wisdom, the medieval times developed theoretical and practical strategies to dominate the environment with technical innovation and theoretical simulation. The enlightenment takes this process further. While scholars in a paradigm that we now can identify as "enlightened vitalism"[18] – between the mechanistic and the romantic period – have sought a synthetic understanding of vital forces in a sensitive way, modern epistemology has set Creation free from all kinds of ideological limitations other than the ultimate concern for accumulating insight for profits. As Maxwell rightly formulates in a recent mail, "it is hardly too much to say that scientific and technological research have made possible all our current problems."[19]

Consequently, modern science and technology can also be interpreted as instruments for salvation in a religious key, even if these are clothed in the dress of postulated secular value neutrality.[20] Climate research, as it is located in environmental science, breaks radically with such a code of science due to an alternative understanding of its normativity. While ordinary natural science seeks insights for the sake of human (often particular) interests, climate and environmental science seeks insights for the benefit of the earth itself, which of course also includes the best for all of its inhabitants in an equal way. The latest approach to climate geopolitics expresses this hopefully in the equal right to emissions necessary for the survival of every global citizen.[21] What however is the best for all, and how it can be achieved is a question for a practical discourse and it should include the participation of all that are affected included non human beings. Global institutions are unfortunately not offering the political institutionalised space and culture for such a dialogue. How should we, however, formulate a consensus about a good life for all and a consensus that acknowledges the intrinsic value of the Earth system in itself as well as the Earth our home? Could a consensus among religious institutions and communities about not "climate as religion" but "climate change as a call to spiritual change" assist the establishment of such a necessary arena? Could it contribute to a deeper fabrication of the meaning of life than economic management of the uncertainties of climate and nature?

Environmental change changes religion

A second theme, worthy of further investigation, is how climate change also implies a change of religion. In several theories, scholars of religion, for example in the anthropology of religion, have analysed how religious modes of perceiving, acting and thinking are dependent on human ecology. The culture of hunters and gatherers for example is connected to specific modes of religious symbolics, where patterns of belief are related not directly to local geography but to the cultural pattern of surviving in a specific environment. Classical Mayan culture and religion, to give another example, interconnects at its heart hydrolatry and urbanisation, and regards its own built empire as a physical expression of the divine powers of life located at the interface of astronomic and subterranean water streams. The story of the garden of Eden in the Hebrew bible expresses the passage from a hunting and gathering community to an agrarian style of life, where the settlement took place after having lost the original good state of life with all the conflicts therewithin.

Historically, environmental conditions are interacting with what we could call a "Sacred Geography" or "Geosophy" implicit in small as well as in big religious traditions. The question is how the ongoing monitoring and discursive construction of climatic change affects the religious fabrication of what it means to be a human person in late modernity. As far as we know from, for example Gaian spirituality, the simple insight that the Earth represents one complex, open and to most of its functions limited system, has catalyzed the construction of holistic Earth religion in a tremendous way.[22] Science does not only investigate nature for purposes, but it also fertilizes alternative religious images of it.

Similar interactions can be observed between new religious processes and their often very selectively chosen pieces of scientific information. Environmental movements are, as Gary Gardener shows in a well-distributed article, deeply driven by spiritual motivations and they increasingly seek, as do politicians, alliances with religious communities and ideologies. "Invoking the spirits" has become a common method. The dynamics of exchange between a religious and scientific fabrication of meaning seems to enter a new state of what we can call "ritualisation". It is obviously exactly this new interchange of normative science and religion that is at the forefront of attacks by scientists, religious right wings and others who have an interest to play down the state of the climate emergency. Another strategy can be found in corporations which rhetorically seek to transform their public image in accordance with the new green religious consciousness fuelled by scientific information. The choice of one company's product then guarantees to contribute to the health of planet Earth. Earth religiosity is turned into a commodity for making economic decisions on the market, where climate change can also be treated as a source of profit making.

The central question for a research process about how changing environments change religion is to explore how academic and public discursive constructions of what environmental change is about are affecting the construction of religious moods, perceptions, ideologies and practises. But research should also turn the other way around. How are religious processes for a deeper sympathy and synergy with the earth system, or let us call it "late modern earth religions", affecting the design and distribution of public discourse? How does the emerging "cosmopolitan" world citizen identity[23] affect the experimentation with practical solutions and the Cultural Revolution which it demands of populations? How are science and religion fabricating a new meaning of life with the planet in a sublime collaboration?

It is obvious in this context how different images of reality are constructed in the rich nations of the North compared to the poorer of the South. A recent survey, which was conducted by the Lutheran World Federation in 2008, investigated how Christian believers in different world regions perceive and interpret environmental change in a spiritual key.[24] Many local people, who could clearly observe the critical changes in their environmental conditions, used to interpret these as either punishments from God or as signs for God's withdrawal from their life worlds, or sometimes even as signs of impending apocalyse. In all cases, Christian belief and theology did not help to encourage scientific perspectives and to question and transform traditional practices of human ecology or the global world system. Rather it worked as a hindrance to the creative and critical adaptation and demand for responsible global mitigation.

Theology must not, however, necessarily function in this way but it can also be turned into a constructive and critical tool for change, as the same survey shows. The challenge to theologians and believers and spiritual leaders in different faith communities has become clearer and more obvious as a first priority agenda. The Manifesto from the Interfaith Climate Summit in Uppsala 2008 for example states the urgency of a spiritual and cultural change, and it emphasizes clearly the significance of faith communities and especially the multitude of religions in this context.[25]

Climate change reveals clearly the historical change of relations between humans and their environments. We know that the last 10.000 years of human civilisation have been possible only because of the stabilisation of climatic conditions for human survival, and one can wonder whether we are now entering the end of this period. The conclusions from the IPCC reports are obvious enough. Without mitigation and adaptation to new conditions, civilisation, as we know it today, cannot survive. Neither can globalised business or the present economic world system. The challenges to faith communities in this new consciousness, also in the power holding bodies, is to resist the temptation to reduce climate change discourse only with the intention of stabilising global capitalism, a driving force behind the Stern report that simply deals with climate change as a question of financial costs, and blends with it an already ongoing victimisation of ecologies of Life.

In spite of enormous scientific and technical innovations, the managers of modern society have lost not only their control but also their skill to imagine what climate does to us at present. They strive to replace uncertainty with new certainties through old empiricist methods. Their vision is a better even more sublime so-called "management of the planet", a vision that seems to me poor, dangerous and in the long run contradictory to the need to develop an alternative agency of nature. Such an alternative would need to be grounded in the spiritual awareness of life as a gift, of nature as a complex ecology with intrinsic value, and an all-embracing space that should be rather interpreted in the frame of metaphors like home, garden and body than as machine, system, or marketplace. Religious traditions hold a wide range of metaphors for life, nature and space which could contribute a lot to the construction of modes of perception, concepts and new theories, even in science. Metaphors and images of life are not just a simple playing with words, but they influence – through language games (Wittgenstein) in words about worlds – our inner images, perceptions and concepts of reality in an elementary way.

Metaphors and imaginations are furthermore, as we know in detail, invisible driving forces behind every theory that has so far emerged in science.[26] The notion of evolution for example, central in biology, was unconsciously transferred by Darwin into his theory of life from the situation of the liturgical unrolling (evolvere) of the holy books in the Jewish synagogue. No wonder that molecular biology today tries to decipher the so called genetic "code" for life, and remains in the image of letters inscribed (by whom?) into the "book of nature" (Augustine), and, by the way, through this method fully misunderstands the function of genes, which recently have been analysed not at all as stable as an alphabet but as much more dynamic movers and "traders", an image more in accordance with late modern experiences of flux, networking and rapid exchange. Words and also images never loose their power over us, not even in science. "In the beginning is the Icon", and as John made clear in his gospel also the Logos.[27]

Another contribution from religion to climate research is the questioning of the strive for reducing and eliminating uncertainty. Not every certainty is good, and not every uncertainty is dangerous. What kind of uncertainty leads us to deep wisdom about life, and what kind of uncertainty should we try to reduce? Could the normal scientific intention to reduce any uncertainty in weather prognosis and climate models lead to a dangerous attitude and practice of simple managing? Or could it assist in developing sensitive attitudes of awareness and learning about the unpredictable dimensions of nature, in a similar way to which indigenous hunters can teach us?

Religions are world champions in dealing with uncertainty and a spiritually differentiated use of uncertainty would be highly necessary also within the limits of empirical science. Scientists can hereby learn from the historical failures of theologians in antiquity and medieval times. If, on the one hand, theologians tried to make certain prognoses about God's being and acting in the world, they would violate the mystery of the Triune in a way that would destroy all kinds of faith. If, they, on the other hand, avoided interpreting God in the modern world, they would only mystify and spread dust around the places and situations where the Spirit acts, liberates and gives new life. The balance of certainty and uncertainty is a necessary deep dimension of spirituality in general, and theologians such as Gregory of Nyssa and Gregory of Nazianz have framed the so-called limits of scientific consciousness in epistemological theory about apophaticism already by the Eastern Patristic age in the fourth century.[28] One should be, apophatically aware, silent about God's essence about which man cannot know anything, but speak positively about his/her actions in Creation, which we can physically experience. The same is true also for nature. Nature in general should not be forced into one single frame of scientific theory but its revelations and many modes of being and fading away should be reflected in concepts characterised by sensibility, awareness and spiritual respect.

Environmental change leads human beings in this regard to the challenge to reflect their own position and identity in and with and for nature anew. How are we located in the "Weltinnenraum" (Rilke)? While the early phase of climate research still tried to determine the uncertain and to make postulates about certainty of development, it nowadays seems much more important to point to the direction where we need to accept and understand, and also theoretically to describe, the intrinsic value of uncertainty implicit in earth systems. The distance between science and religion and culture shrinks in such approaches, even if very few models of a constructive cooperation of science and religion have been initiated so far.

Urbanisation, Mobility and Acceleration – Lived Religion in Lived Space

One of the driving forces behind civilisation is, as geographer Edward Soja has shown, the development of city space as it had begun already more than 10.000 years ago, for example in Catal Hüyük.[29] The accelerating process of urbanisation is now turning the whole planet into one single "postmetropolis". This affects the reshaping of landscapes and regions all over and a new recently formulated research agenda concentrates on how religion is affected and interconnected with urbanisation. One can wonder whether it is this deep urban stimulus that represents one of the central driving forces behind and within the historical process of an evermore destructive world civilisation overheating its own atmosphere.

With regard to climate change city space promotes and accelerates at present fuel driven and profit oriented industrialism and motorised mobility which offers one of the largest sources for emissions. Hypermobility has reached a state of unpredictability that threatens the whole system of social and environmental planning and even democracy in its roots. The French philosopher Paul Virilio characterises the present state of culture as a standstill caused by increasing acceleration which German sociologist Hartmut Rosa has described as a central attribute of modernity.[30]

Religious life views can in this sense be studied in the conflict about mobility which is so typical of modernity. The dispute in the beginning of the 20th century between futurist and Dadaist artists offers a microcosm of this controversy which also shapes climate discourse. While futurists were glorifying speed, progress and technical innovation and were enchanted by its potential to change urban life worlds, the Dadaists objected to the fascist glorification of strength, progress and the technocratic attitude towards natural life. While the one depicted a world of change and construction, the other staged a world of waste and destruction.

2 Giacomo Balla, Auto en Course, Etude de Vitesse, 1912, oil on canvas, 55,6 x 68,9 cm, The Museum of Modern Art, New York, © Giacomo Balla/BONO, Oslo 2010

Religious attitudes can, as we have seen, take two directions. Either they will follow the futurist glorification and politics of speed and progress, or they will cultivate the Dadaist desire for an alternative mode of organizing city life. A central question, also for the challenge to apply climate change conclusions to an alternative life style is how a city can be turned into a habitable place and what religion can contribute. What are the qualitative criteria for habitability and how can these be designed with regard to what we know about emission and energy house holding?[31]

The perception of the built environment is in such a view at the core, and spiritual and immaterial values as well as criteria of ecologic justice need to be reflected as significant driving forces for an urban transformation. I have earlier called such an ap-

3 Hannah Höch, Mechanischer Garten, 1920, watercolour, 73,5 x 46,4 cm, The Museum of Modern Art, New York, © Hannah Höch/BONO, Oslo 2010

proach for "aesth/ethics", which implies the integration of ethics in our bodily being. Modes of morally acting and moralities need to be integrated in our bodily modes of perceiving each other in surroundings. Changing climate offers in this sense a central potential as the knowledge and interpretation of climate needs to be in harmony with our bodily perceptions of what climate does to us.

Looking at the latest formulations of the EU and their working programmes' call for research on climate change we find a total lack of such an acknowledgment of the human dimension of the field of problems.[32] In the EU horizon one only needs to monitor and analyse the phenomena, environmental technology innovation and suggestions for policy designing. Also scientists usually place structural solutions, mostly economic tools, over human and cultural skills, which they regard as ideologically helpful but not really important.[33] The image of humanity in such perspectives seems to me highly problematic and undemocratic. What we can call the human dimension of environmental science and climate change research, as it has been unfolded here in its spiritual and aesthetical shape, is not even existent in common approaches to research. It is perhaps no wonder why reductionist explanations will only produce fragmentary solutions that can be used for power play cosmetics, while the real originating source of the problems, human beings themselves, are mainly neglected? Climate science obviously still lacks a social power analysis, which could help to differentiate and for example differ between geopolitical engineering of transnational corporations on the one hand and the development of local populations on the other. Without such a self-critical power analysis climate science is threatened to remain imprisoned in existent power constellations.

My argument for a greater acknowledgment of local populations' skills and especially perception in research perspectives aims to rebalance in the light of these difficulties, and it follows the well-known anthropologist Tim Ingold who prioritises the perception of the environment over the knowledge about it and the action in it. Indigenous populations and cultures are hereby radically different from modern cultures and they offer interesting skills that might be reconsidered for our common future. While the hunter interprets his/her surrounding with bodily skills and in the frame of a spiritual cosmology where life takes place in a given space, s/he is extremely dependent on using aesthetical skills for surviving.

Anthropologist Roy R. Rappaport has furthermore shown in his ritual studies how environmental regulations could take place as efficient practices in the ritualisation of a culture's agency with natural resources.[34] It is however controversial in Ritual studies, how this works and whether rituals *should* function as environmental regulators or in fact are *doing* it. If ritualisation is a basic social act, and I think it is, it represents a "cosmo-topic" practice where the human cosmos is established performatively as a "cosmo-sized place", a universe at a local place.[35] How are ritualisations of climatic change taking place?

For a detailed study of how the meaning of changing climate is culturally handled, Timothy B. Leduc has offered excellent observations and reflections about the Arctic Inuit understanding of weather as "Sila", which is perceived as a spiritual power, and as Leduc rightly claims, necessary for climate research and social justice to respect and somehow to integrate in its cross-cultural interdisciplinarity.[36] His chapter in this book deepens this even more. Cultural spiritual wisdom based on bodily perceptions is needed to lead research to a broader frame.

Observations like these compel us to ask how aesthetical skills can be related to empirical and cultural studies. How can the bodily perception of our environment and the ritualisation of cosmosized places be developed as a social strategy for change? In short: how could science inform our eye and religion our soul to perceive a changing environment and ourselves as a part of it in an alternative way? How could a deep integration of perception and knowledge lead to wisdom?

Modern societies are in opposition to indigenous small-scale cultures characterised by an opposite pattern, that is the distancing from nature and the cutting off from feedback processes which are actually taking place in nature. A simple meal for example involves geographical and local conditions for agriculture, as well as fuel and money driven transport and distribution systems before they reach a kitchen where again energy is used to prepare them for the table. We usually do not get even a glimpse of the ecological complexity beyond and implicit in a simple lunch. Feedback circles, as the reknowed Gregory Bateson emphasized them, are structurally expelled from our consciousness. We act and think in a universe which we have designed in a way that robs us of information and perceptions of how things are interconnected. The new religiosity's commitment to holistic and universal values, where connectedness is at the core, can only offer poor substitutes for this lack of global belonging.

If one follows Jakob von Uexküll's famous circle of function (Funktionskreis), not only the single organism's life world is connected in a bi-directional way between the "Merkwelt" and "Wirkwelt", but climatic change appears as a gigantic challenge to embed and visualise the flows of this circle of function between our perception and our action on a global and local scale. Global space with all its local places need to be revised as a cosmos of synergies, where the symphony of things moves back into our consciousness.[37] The complexity of climate studies and the phenomena of interconnectedness offer a new source to catalyze such a changing worldview of a lived space. Religion, regarded (in Kant's words for art), as a force of productive imagination (produktive Einbildungskraft) is crucial in the re-imaging of our environment.

Following philosopher Henri Lefebvre, Soja has made a theoretically useful distinction between three types of space, physical, imagined and lived space.[38] The concept of "lived religion", as it has been developed for the phenomenological study of religious practices in place and space, coincides well with this concept and it offers an exciting tool to an approach of studying climatically caused religious change in the city space of postmetropolis.[39] The spiritual and aesthetical skill to identify one's life in built environments as a meaningful life is a strongly religiously shaped process of identification.

Rappaport has in his latest book described religion as a "fabrication of meaning". He differentiates between a scientific and a religious mode of approaching reality and argues for a necessary synthesis between the two.[40] For him the elementary tension of human life lies between the fabrication of meaning with regard to our environment and "the epistemologies of discovery" which aim to explain the laws of this environment.[41] Applying his reflection on climate change we can state a conflict between the rational monitoring and explaining of a changing global and local environment on the one hand and the fabrication of meaning of this change on the other. Both need to be worked out, but scientific explanation can and should not replace the ritualised religious fabrication. The tension and dissonance of law and meaning need to be acknowledged in its depth. The relevance of such an approach has as far as I know not been explored in climate change.

My hypothesis is that our construction of meaning in the discourse about dangerous environmental change in itself offers a strong and important chance to manifest such a synthesis of scientific explanation and religious meaning. This does not mean that everything in science that fuels the religious attitude to life is good, or that every scientific analysis needs to be transformed into a policy-generating strategy where citizens' behaviours are simply steered by designed information for action. It construes the citizen him/herself as the main agent of perception, knowledge, wisdom and meaning fabrication, and therefore as the agent of a change of life styles and patterns of production and consumption. The contribution of Religious studies in this context is to guarantee that the integrity of the believer is understood as the main source for social transformation and that s/he is not simply treated as an object for the design of scientific information and socioeconomic management policy.

An important source of strength in this context can be found in the process of remembrance. Especially the remembrance of the suffering previous generations offers a crucial potential for the development of a sound transformation of life styles and worldviews. As we know climate change is not just a coming threat but it has already resulted in a large number of victims among human beings, species and ecosystems all over the planet. Again, the poorest nations must carry the greatest sufferings. How do we deal with the memory of these suffering neighbours and co-creatures? And how do we deal with projections of future sufferings? Ethical decisions are not simply made by moral arguments in theory according to universal principles. They are always embedded in local perceptions and practices in social and natural environments.

Theologian Johann Baptist Metz reflects in a famous essay about the significance of the remembrance of suffering.[42] The remembrance of suffering is not just a task but that it is necessary in order to find and apply the processes of true liberation from powers of teleological and technological thinking. As I have shown in another context, the demand to integrate the remembrance of suffering is crucial for the process of urban planning,[43] and one should therefore also explore it within the framework of climatic change. How could remembrance of the climate-caused suffering of earth inhabitants contribute to the new ritualisations of environmental planning and to fabrications of the meaning of lived space and lived religion in a climatically changing world?

Such a perspective transcends of course the sphere of academic research by taking us to the inner life of faith congregations. It leads them to ask: How does the Holy Spirit take place in climatic change? How do the communities of believers interact with a liberating God in lived space? How do they transform built environments into habitable space, flourishing in ecologic justice both for humans and for others?

Even if answers need to be worked out in pastoral theology, academic theology can assist to point out the specific qualities of contributions, which can be found far beyond simple ethical reasoning from biblical principles in experiential spiritual, ritual and systematic-historical processes, which only religious communities are capable of experimenting with.

THE RECIPROCITY OF ENVIRONMENTAL CHANGE AND RELIGION – TOWARDS A RESEARCH AGENDA

How could one develop an agenda for integrated research on climate change and religion where the reciprocity between them is at the fore? As we have seen thusfar, not only the production of anthropogenic climate change but also the construction of the discourse about it implies a crucial challenge for a self-critical reflexivity which Religious studies and Theology could help to deepen. The changing interactions of science and religion are in this context offering a genuine challenge to a truly transdisciplinary environmental science, which operates openly and transparently with its implicit values, worldviews and commitments.

In the following I will list only some of the approaches to religion, as they figure in the academic field, and depict their potential contributions to a broader research field. All these could be broken down into concrete projects or subfields of research. Preliminarily I would depart from the following twelve perspectives for the design of a future research programme on climate change and religion.

a) Religious individuals, practices, traditions, and institutions might contribute to the production of environmental problems as well to their solutions. Religiosity works in a destructive as well as in a constructive key. The study of the history of religions and spirituality shows how this dilemma and the *ambiguity of religious attitudes to nature* has been a main theme in religious institutions as well as in several modern western critiques of religion. Also with regard to global change, religion could serve as a destructive as well as constructive force.
b) The *sociological study of environmentalism* has strongly impacted on the discourse about the Western society in general, its colonisation, modernisation, globalisa-

tion, and secularisation and most recently, its re-sacralisation. Connections between these two discourses in social sciences on environmentalism and religiosity have shown how fruitful and enriching such a synthesis might be for the future development of social sciences in general. *Gender Studies*, which, since the mid-90s has focussed on the field of ecofeminism, make even more evident how important a study of the interconnectedness of Gender, Nature, and Religion will be for the understanding and solution of increasing problems in the Western self-identity. *The environmental movement* in itself could be studied as a religious movement on the one hand; on the other hand, religious institutions and processes could be analysed as a social movement in the context of a green civilisation critique and the population's desire for another life beyond the powers of modernity.[44] A sociological discourse shows how the changing connection of science and religion catalyzed by climate change discourse offers a further crucial element to the so-called resacralisation of late modernity.

c) A third part of a sociological argument is the affectedness by and the adaptation of the *world religions,* their leaders, institutions, and practices by the emergence of an ecological consciousness, politics, culture, marketing, etc. Sociologists have analysed this field in detail and we could even speak of an environmental paradigm in world religions entering the discourse, scenes, and social practices of ecology and sustainability. A comparative study of how religion works in different places, traditions and contexts with regard to ongoing environmental change would be able to deliver valuable insights into the dynamics in the human ecological triangle of subjective, sociocultural and physical dimensions of human ecology in change.

d) With regard to the social transformation of modernity, the emergence of so-called *new religiosity/spirituality* also represents a strong empirical phenomenon that has been intensively reflected in the last years. Climate change is in this context seen as evidence for the need for a new spiritual attitude to the planet as a holistic earth in itself.

e) A further approach can be found in *cultural studies* that do work with several concepts of religion and religiosity.[45] Either religion is regarded in its autonomy as a unique force of cultural identity and practice or it is used as a tool in a functionalist or constructivist way: thus religiosity in the space between ideas and praxis, between worldview and ethos, as Clifford Geertz puts it, is a main theme in the cultural studies' contributions to human ecology. As cultural images and practices are changing and affected by the climate change discourse also cultural and religious identities and images of reality are changing. A crucial contribution from an anthropology of religion approach is, as we have seen above, the exploration of skills of perception and ritualisations as strong elements of fabricating meaning for life.

f) *Economically* the impact of environmental movements on the pattern of production, distribution and consumption of goods and services has been increasing and in many places it has caused earthquakes in the landscapes of financial and industrial late capitalism. Consumers have experienced a new form of power accumulation with regard to producers and markets, and the greening of business management is obvious. How are the underlying cultural and civil-religious dimensions of these kinds of developments to be understood? How does cosmopolitan earth religiosity affect economic behaviours?
g) *Politically* the alliances of religious and political institutions have deepened and developed since the 80s in a way that cannot be understood by conventional secularisation theories. Leaderships have cooperated in regional, national, continental and global arenas. In the alliances with social base movements, NGOs, and religious communities, we can observe large- and small-scale activity and a multitude of political processes involving believers, faith communities, and other agents in politically and NGO-defined political networks. Is there a stronger influence of religiously constructed worldviews and values on politics? How to explore an increasing influence of images of the Sacred[46] on the politics of nature and the climate?
h) Detailed studies in the Sociology of Religion show how new religious processes of identification in general operate with the help of concepts of nature, life, and bodily being. In fact the question is whether we could talk about *a new emerging civil religion* characterised by what scholars have started to call "ecological spirituality", and how it interacts with climate and religion change. The metaphor of the "global village" and its success in ruling ideologies shows clearly how the understanding of sovereignty, political authority, and communal identity has gone through a shift from a nation-based to a global mode of identity construction. It seems that it no longer is national identity that is at the heart of the human person, but his/her localisation as a citizen in the global horizon. An important question arises from this: Is a conception of political community and citizenship mainly built on the central basis of cosmopolitan or terrapolitan identity?[47] Is the global village mainly understood as the Earth itself? The views of nature, life, and Earth including climate seem to be at the heart of the changing self-understanding of persons and citizens in political and moral communities.[48]
i) Environmental problems in the interaction of nature and man/woman are one of the best test cases for how civil development to a more peaceful, sustainable, and just world society could be achieved. *Environmental studies* should depart from the general insight that all kinds of human problems are to be dealt with by means of the model of the human ecological triangle: that means that they are problems in the space between subjective, sociocultural, and environmental dimensions. Environmental problems are seldom only technical or political problems: they always

indicate a deeper wound in the areas of the human soul/body, the sociocultural and natural body of the surroundings of human beings. Lost worlds and lost words are related, a phrase coined in geography. The loss of human words is a catalyst for losing the worlds surrounding us. Loosing our surroundings means loosing our language.

j) Technologically the status and significance of *environmental technology* needs to be clarified. Regarded in the light of Religious studies, technology is often marketed as a tool of problem solution and even salvation.[49] It generates illusions of easy and quick solutions without greater costs. How are cultural constructions of technical fix developed? How is the role and substance of technology in climate change discourses described and how can it be critically reflected? How are ideologies of technology management developed as doctrines of salvation and domination rather than tools and elements in a more complex theory of what a changing environment and the enhancing of life demands?

k) Spatially, last but not least, Religious studies need to develop a reflection of lived religion in lived space as the distribution of urbanisation in all world regions is accelerating violently at present. What is a habitable place to live in and how can the criteria of a flourishing and ecojust life world be promoted by *third space research* that explores the tools for changing urban planning with regard to new climatic affordances? How can the remembrance of the suffering and victims of climate change direct the living generations for the sake of the coming ones?

l) A crucial challenge to the reorganisation of *science* itself is the need to create a new balance of faculties in order to develop multifaceted tools to explore and encounter climatic change. The widening of environmental science in the 90s from natural science, technology and medicine by also including sociological approaches has to be completed with studies on *the human dimension*. Otherwise environmental problems cannot be approached satisfyingly in analogy to the complexity of the problems themselves. Transdisciplinarity is in itself a highly explosive topic with regard to the future research policy of national and global decision makers. Without this approach, higher education and research will continue to produce more problems than it will ever be able to understand and solve. Science and technology become in Marx' term the perfect commodity, if they do not limit themselves with regard to a limited Earth system. In fact, academic, non-environmentally enlightened science should be regarded as a political, economical, and cultural problem and threat in itself.

An alternative approach with such a vision creates new conditions for humanities to enter and develop transdisciplinary alliances. The discipline of Religious studies and Theology could in itself be regarded as a microcosm of humanities and cultural studies, which makes it specifically interesting to link scholars of religion to a new and truly *transdisciplinary environmental science* to come. Climate, and the committed environmental change discourse, would then not simply turn into a "climate as religion" position but search for creative interactions of different religious and cultural wisdom with self-critical research in empirical, hermeneutical and aesth/ethical academic spheres. In collaboration such a study of religion-change *in* environment-change would open up a broad wisdom inquiry where science is just one of several birth-giving agents in the majeutics of new insights.

NOTES

1 Lawson.

2 <http://www.globalwarminghysteria.com/environmentalism-as-religion-/>, 29 November 2007.

3 McKenna.

4 Some of many examples in the following. The World Council of Churches (WCC) has worked on climate issues since 1990 and, for example, published the document *The Atmosphere as Global Commons: Responsible Caring and Equitable Sharing*, 2000. <http://www.wcc-coe.org/wcc/what/jpc/cop6-e.html>, 31 October 2007. A working group for climate change has worked since 2001 in the European Churches Environmental network (ECEN), and a programme for climate justice is developing in the European churches, <http://www.ecen.org/cms/index.php?page=climatechange>, 31 October 2007. In the USA the Evangelical Environmental Network adresses global warming "as a new way to love our neighbors", <http://www.creationcare.org/files/global_warming_briefing.pdf>, 31 October 2007. American evangelical leaders published 2006 *Climate Change: An Evangelical Call to Action*, <http://www.npr.org/documents/2006/feb/evangelical/calltoaction.pdf>, 31 October 2007. The Evangelical Church in Germany (EKD) published 1995 its study *Gefährdetes Klima*, <http://www.ekd.de/EKD-Texte/44652.html>, 31 October 2007, and its chair W. Huber has adressed decision makers in his *It is not too late to respond to climate change*, May 2007, <http://www.ekd.de/download/ekd_texte__89_engl(1).pdf>, 31 October 2007.

5 Cf. for example Daly and Cobb, and Cobb.

6 Weizsäcker, 3ff.

7 Cf. Curtin.

8 Des Jardins, 5.

9 World Council of Churches (1998) and (2002). Cf. also Steigerwald.

10 *Zukunftsfähiges Deutschland*. The Wuppertal Institute's social and epistemological model is in my view poor, as it regards the "experts" as producers of knowledge and the politicians and social movements as those who simply need to apply the right insights in policies and lifestyles.

11 EU Commission for research, Draft for a Work Programme "2008 Cooperation, Theme 6, Environment (including climate change)", October 2007.

12 Northcott, 7, 118.

13 Næss, 37-40.

14 Maxwell.

15 Stern.
16 Cf. Jinwol and Tucker and Williams.
17 Gregory of Nazianzus, Oratio 38.13, 16.5, 6.14. Bergmann (2005a), 109.
18 Reill.
19 Nicholas Maxwell, Are philosophers responsible for global warming? Email letter to the mailing list <PHILOS-L@Liverpool.ac.uk> 13 October 2007.
20 Cf. Bergmann (2007).
21 Equal rights to emissions for every global citizen have been proposed by German chancellor Angela Merkel in September 2007. Such a proposal would, however, need to acknowledge the difference between luxury and ordinary life emissions. The principle of equal rights is also applied in the Norwegian engineers' report *Ethics of Climate Change: Exploring the principle of equal emission rights* by Inge Johansen for the Norwegian Academy of Technological Sciences (NTVA), Trondheim 2007.
22 A representative example is found in Primavesi (2000). On earth spirituality see also Bergmann (2005b).
23 Cf. Deudney.
24 Cf. Bloomquist.
25 *Hope for the Future!*
26 Blumenberg, 19.
27 Cf. Bergmann (2009), chapter I.
28 Cf. Bergmann (2005a), 339ff.
29 Soja (2000), 35.
30 Virilio, Rosa. Cf. Bergmann (2008a).
31 Cf. Gorringe and Bonsdorff.
32 Cf. above note 11.
33 Also a committed and politically very influential climate scientist like Christian Azar represents such a fatal split between socio-engineering tools and human skills (Johan Falk, "Klimatguru med rätt att flyga", in: *Forskning & Framsteg* 1, 2009, 18-21).
34 Rappaport (1968). Cf. Grimes.
35 Grimes, 44.
36 Leduc.
37 Cf. Bergmann (2006).
38 Soja (2004).
39 Cf. Bergmann (2008b).
40 Rappaport (1999).
41 Rappaport (1999), chapter 14.
42 Metz.
43 Bergmann (2008c).
44 A very strong argument has been developed in detail by the influential World Watch Institute's director Gary Gardener who makes a strong plea for the political and cultural alignment of environmentalists and believers of all kinds of traditions. Gardner's list of already documented social processes and the ongoing reception of his essay makes an argument and empirical fact in itself that shows the depth and potentials in the outlined field.
45 Cf. Witoszek-FitzPatrick.

46 Szerszynski (2005a).
47 Cf. Deudney and Scott.
48 For a differentiated systems approach on responsible global citizenship with great significance for Europan developments, see Blackmore and Smith. The prominent sociologist Peter Beyer analyses the ongoing transformation of values and practises with his "re-sacralisation" notion. This implies that late modernity is no longer deepening its secularisation, but is characterised by the emergence of new public religions entering public discourses, in order to state the credibility and the ethical and practical plausibility of their belief system with regard to normative and explosive themes – for example, in the discourse about our common sustainable future and environment.
49 Bergmann (2007). Cf. Szerszynski (2005b).

References

Bergmann, Sigurd (2005a), *Creation Set Free: The Spirit as Liberator of Nature*, (Sacra Doctrina: Christian Theology for a Postmodern Age 4), Grand Rapids Mich.: Eerdmans, (German ed. *Geist, der Natur befreit*, Mainz: Grünewald 1995, Russian ed. Arkhangelsk: Arkhangelsk University Press 1999).

— (2005b), "Space and Justice in Eco-Spirituality", in: Vasilios N. Makrides and Jörg Rüpke (eds.), *Religionen im Konflikt: Vom Bürgerkrieg über Ökogewalt bis zur Gewalterinnerung im Ritual*, Münster: Aschendorff, 212-225.

— (ed.) (2005c), *Architecture, Aesth/Ethics and Religion*, Frankfurt/M./London: IKO-Verlag für interkulturelle Kommunikation.

— (2006), "Atmospheres of Synergy: Towards an Eco-Theological Aesth/Ethics of Space", *Ecotheology: The Journal of Religion, Nature and the Environment*, Vol. 11, 3, 327-357.

— (2007), "Technology as Salvation? Critical Perspectives from an Aesth/Ethics of the Spirit", *European Journal for Science and Theology*, Vol. 3, 4, 5-19.

— (2008a), "The Beauty of Speed or the Discovery of Slowness – Why Do We Need to Rethink Mobility?", in: Bergmann and Sager, 13-24.

— (2008b), "Lived Religion in Lived Space", in: Heinz Streib, Astrid Dinter and Kerstin Söderblom (eds.), *Lived Religion: Conceptual, Empirical and Practical-Theological Approaches – Essays in Honor of Hans-Günter Heimbrock*, Leiden/Boston: Brill, 197-209.

— (2008c), "Making Oneself at Home in Environments of Urban Amnesia: Religion and Theology in City Space", *International Journal of Public Theology*, Vol. 2, 70-97.

— (2009), *In the Beginning Is the Icon: A liberative theology of images, arts and culture*, London: Equinox.

Bergmann, Sigurd and Kim, Yong-Bock (eds.) (2009), *Religion, Ecology and Gender: East-West Perspectives*, (Studies in Religion and the Environment 1), Münster/Berlin/Zürich/Vienna/London: LIT.

Bergmann, Sigurd and Sager, Tore (eds.) (2008), *The Ethics of Mobilities: Rethinking Place, Exclusion, Freedom and Environment*, Aldershot/Burlington: Ashgate.

Beyer, Peter (1994), *Religion and Globalization*, London: Sage.

Blackmore, Chris and Smith, John (2002), "Living with the Big Picture: A Systems Approach to Citizenship of a Complex Planet", in: Nigel Dower and John Wiliams (eds.), *Global Citizenship: A Critical Reader*, Edinburgh: Edinburgh University Press, 201-212.

Bloomquist, Karen L. (ed.) (2009), *God, Creation and Climate Change: Spiritual and Ethical Perspectives*, (LWF Studies 2/2009), Geneva: The Lutheran World Federation.

Blumenberg, Hans (1993), *Die Lesbarkeit der Welt*, Frankfurt am Main: Suhrkamp 3. ed. 1993 (1981).

Bonsdorff, Pauline von (2005), "Habitability as a Deep Aesthetic Value", in: Bergmann (2005c), 114-130

Cobb Jr., John B. (1992), *Sustainability: Economics, Ecology & Justice,* Maryknoll: Orbis.

Curtin, Deane (2002), "Ecological Citizenship", in: Engin F. Isin and Bryan S. Turner (eds.), *Handbook of Citizenship Studies*, London: Sage, 293-304.

Daly, Herman and Cobb Jr., John B. (1994), *For the Common Good: Redirecting the Economy Toward Community, Environment, and a Sustainable Future,* rev.ed. Boston: Beacon Press (1989).

Des Jardins, Joseph R. (1997), *Environmental Ethics: An Introduction to Environmental Philosophy,* 2. ed. Belmont: Wadsworth.

Deudney, Daniel (1998), "Global Village Sovereignty: Intergenerational Sovereign Publics, Federal-Republican Earth Constitutions, and Planetary Identities", in: Karen T. Litfin (ed.), *The Greening of Sovereignty in World Politics*, Cambridge Mass.: MIT, 299-325.

Gardner, Gary (2002), *Invoking the Spirit: Religion and Spirituality in the Quest for a Sustainable World,* Worldwatch Paper 164, Washington.

Gorringe, Timothy J. (2002), *A Theology of the Built Environment: Justice, Empowerment, Redemption*, Cambridge: Cambridge University Press.

Grimes, Ronald (2003), "Ritual Theory and the Environment", *The Sociological Review*, Vol. 51, 2, 31-45.

Hope for the Future! The Uppsala Interfaith Climate Manifesto 2008, Faith traditions addressing Global Warming, (2008), Uppsala: The Church of Sweden, <http://www.svenskakyrkan.se/default.aspx?di=173302>, 26 January 2009.

Jinwol, "Seon Experience for Ecological Awakening", in: Bergmann and Kim, 131-145.

Maxwell, Nicholas (1984), *From Knowledge to Wisdom: A revolution in the aims and methods of science*, Oxford: Blackwell.

McKenna, Phil (2007), "Climate change unites science and religion", *New Scientist*, 17 January 2007.

Metz, Johann Baptist (1972), "Erinnerung des Leidens als Kritik eines teleologisch-technologischen Zukunftsbegriffs", *Evangelische Theologie,* Vol. 32, 338-352.

Northcott, Michael S. (2007), *A Moral Climate: The ethics of global warming*, London: Darton, Longman and Todd.

Næss, Arne (1989), *Ecology, Community, and Lifestyle: Outline of an Ecosophy,* Cambridge: Cambridge University Press, (shortened and revised edition of *Økologi, samfunn og livsstil: Utkast til en økosofi,* Oslo: Universitetsforlaget 1974).

Lawson, Nigel (2006), *The Economics and Politics of Climate Change: An Appeal to Reason,* A Lecture to the Centre for Policy Studies, 1 November 2006, <http://www.cps.org.uk/cpsfile.asp?id=641>, 29 November 2007.

Leduc, Timothy B. (2007), "Sila dialogues on climate change: Inuit widsom for a cross-cultural interdisciplinarity", *Climatic Change*, Vol. 85, 3-4, 237-250.

Primavesi, Anne (2000), *Sacred Gaia: Holistic Theology and Earth System Science,* London/New York: Routledge (with a foreword by John Lovelock).

Rappaport, Roy A. (1968), *Pigs for the Ancestors: Ritual in the Ecology of a New Guinea People,* New Haven: Yale University Press.
— (1999), *Ritual and Religion in the Making of Humanity*, Cambridge: Cambridge University Press.
Reill, Peter Hanns (2005), *Vitalizing Nature in the Enlightenment*, Berkeley/Los Angeles/London: University of California Press.
Rosa, Hartmut (2005), *Beschleunigung: Die Veränderung der Zeitstrukturen in der Moderne*, Frankfurt/M.: Suhrkamp.
Scott, Peter (2003), *A Political Theology of Nature,* Cambridge: Cambridge University Press.
Soja, Edward W. (2000), *Postmetropolis: Critical Studies of Cities and Regions*, Oxford: Blackwell.
— (2004), *Thirdspace: Journeys to Los Angeles and Other Real-and-Imagined Places*, Malden/Oxford/Carlton: Blackwell, reprinted (1996).
Steigerwald, Jutta (2008), "Walk the talk: Mobility, Climate Justice and the Churches", in: Bergmann and Sager, 241-253.
Stern, Nicholas (2006), *Stern Review on the Economics of Climate Change*, London: HM Treasury.
Szerszynski, Bronislaw (2005a), *Nature, Technology and the Sacred*, Oxford: Blackwell.
— (2005b), "Rethinking the Secular: Science, Technology and Religion Today", *Zygon*, Vol. 40, 4, 813-822.
Tucker, Mary Evelyn and Williams, Duncan Ryuken (eds.) (1997), *Buddhism and Ecology: The interconnection of dharma and deeds*, (Religions of the world and ecology), Cambridge, Mass.: Harvard University Press.
Virilio, Paul (1999), *Fluchtgeschwindigkeit: Essay,* Frankfurt/M.: Fischer (1995).
Weizsäcker, Ernst Ulrich von (1994), *Erdpolitik: Ökologische Realpolitik an der Schwelle zum Jahrhundert der Umwelt,* 4. ed. Darmstadt: WBG, (engl. ed. *Earth Politics*).
Witoszek-FitzPatrick, Nina (1998*), Identity, Community and the Natural World*, a special issue of *World Views*, Vol. 2, 3, (ed. with Andrew Brennan).
World Council of Churches (WCC) (1998), *Mobilität: Perspektiven zukunftsfähiger Mobilität*, Geneva: WCC.
— (2002), *Mobile – but not Driven: Towards equitable and sustainable mobility and transport*, Geneva: WCC.
Zukunftsfähiges Deutschland: Ein Beitrag zu einer global nachhaltigen Entwicklung, (1996), Basel: Bund, Misereor.

Adapting to Climatic and Hydrologic Change

Variegated Functions of Religion

Dieter Gerten

What Western scientists call "ecological processes", like the "water cycle", some native cultures call "water persons" and "relatives".
(Melissa K. Nelson)[1]

Culture and religion in climate and environmental change

The further global climate change proceeds, the more likely the biophysical and ecological conditions around the world will be fundamentally altered. For example, the probability and frequency of extreme weather events (droughts, storms, heat waves, etc.) is projected to increase in most regions; the global sea level is expected to rise significantly; and terrestrial, aquatic, and marine ecosystems are likely to undergo fundamental changes.[2] In addition to the anthropogenic climatic changes and their impacts, humans are increasingly altering the landscapes of the Earth in more direct ways – for example through deforestation for the purpose of food and timber production; through overfishing of oceans; or through exploitation and/or contamination of non-renewable freshwater resources.[3] One major manifestation of the current environmental problems is the emergence of a "global water crisis", that is, the progressive diminution and degradation of regional freshwater resources in view of an ever-increasing world population.[4] As collective human activities have significantly transformed the surface characteristics and the biogeochemical balances of the Earth (particularly since the 1950s), it appears that we have entered a new, quasi-geological era: the so-called "anthropocene".[5]

Importantly, not only are humans to a large extent responsible for the ongoing environmental changes, but also these changes do feedback in multiple ways to human societies and culture. For instance, climate change, and especially extreme events, may result in pronounced loss of human life and properties,[6] and may well induce mass migrations of human populations from some regions.[7] There is also ample evidence that distinct cultural and religious values of individuals and whole societies influence their perception and tolerance of risks as well as their capacity to cope with environmental hazards.[8] Also, the social (including the religious) background of an individual is a prime determinant of his/her environmentally relevant behaviour.[9]

Hence, culture – and religion, especially in non-Western and less secularised societies including indigenous communities – crucially shapes human interactions with their environment and with specific places,[10] and, thus, climatic and other environmental changes significantly affect cultural practices. This calls for a consideration of cultural specifics as an integral part of studies of the impacts of climate change, and adaptation to these impacts. Despite this evidence, the current academic discussion about adaptation to climate change – i.e., the private or public "adjustment in natural or human systems in response to actual or expected climatic stimuli or their effects, which moderates harm or exploits beneficial opportunities"[11] – is centred around the promulgation of technological solutions to reduce societal vulnerability to climate change impacts, and appears to be biased toward economic cost–benefit analyses. Comparably little attention has been paid to the cultural dimensions of climate impacts and adaptation, for example the psychological dimensions of adaptation.[12]

In this chapter, I provide examples of how religious practices are affected by climatic/hydrologic changes, and I investigate the different roles that religion may play in adapting to such changes.[13] The focus is on freshwater-related changes, because a) water has played and continues to play a crucial role in virtually all belief systems and practices from ancient times up to today;[14] b) water systems are increasingly threatened all over the world;[15] and c) the integration of cultural, religious and ethical principles in water studies is quite advanced, not least because there is now a shift in water management towards a new paradigm that no longer treats human culture and belief systems as mere boundary conditions but integrates them in a more holistic view.[16]

In the following sections I discuss selected case studies demonstrating a variety of religious perceptions of, and religiously motivated responses to, changing hydrological conditions. In the concluding discussion I try to synthesise these findings and tackle the question how religious studies could contribute to research on climate change and its (hydrologic) impacts.

Can religion promote sustainable water use?

Water in its various aggregate forms plays a core role in virtually all mythologies and religions. Springs, rivers, lakes and the sea are worshipped as divine manifestations, as the source and bloodstream of all life, and as a medium with healing powers; but at the same time, water is feared for its destructive powers especially during floods. Since religious beliefs underlie the fundamental conceptions of water (and of nature in general) of billions of people around the world, it is likely that many people's habits of water use are at least to some extent influenced by religious considerations.[17] For example, because of the fact that water is regarded as a sacred element in religious texts and rituals, any misuse and overexploitation of freshwater may be regarded as an abuse of the divine (e.g., as a sinful act).[18] It has to be noted, however, that this religious influence on environmentally relevant behaviour is not necessarily a positive one but can be ambiguous and sometimes even adverse, as I will outline further below. Nevertheless, there is a body of evidence that religious considerations – and perhaps even more so the related ethical and moral principles – can promote sustainable and ethically informed means of water use, as expressed e.g. in petitions by indigenous communities such as the Indigenous Peoples Kyoto Water Declaration.[19]

One of many examples for the positive role of religion in freshwater management is the successful restoration of Lake Biwa, one of the world's most ancient and Japan's largest lake, its "Mother Lake".[20] In the 1960s it became obvious that this lake underwent a dramatic decline in water quality and ecological status, which was mainly a consequence of agrochemical pollution. A first wave of environmental laws established around 1970 – which regulated the use of fertilisers, the abandoning of agricultural land in the surrounding area, and the control of effluences – was not sufficient to halt further degradation of the lake's ecological status. It was only the Lake Biwa Ordinance issued in 1979 that reversed the eutrophication trend of the lake.

Importantly, the first step towards this law was the engagement of the local citizens, who recognised "that it was time to shift from a consumer society to a more ecological lifestyle, and resume the ethical principles that were embedded in their ancestral traditions".[21] It appears that the restoration of the lake to its former ecological status would not have been successful if local citizens and authorities had not changed their way of thinking about the lake and about water and nature in general: in fact, the remembrance of the "ancestral traditions" is characterised by the wish for harmonising humans and nature (following ethical and aesthetic principles), for intergenerational solidarity including responsibility for universal heritage, and by the respect for sacred water and land spirits – the *kami* of the animistic Shinto religion that, together with Buddhism, is still prevalent in Japan. In sum, the deeply rooted Shinto-Buddhist worldview stimulated the political will for change and the establishment of a "new" environmental value system, which found its way not only into the local Lake Biwa Ordinance but also into Japan's environmental policy in general.[22]

While examples like this suggest a positive influence of traditional belief systems on how people care for their environment, many people fail to recognise the (sometimes buried) "greening" potential of most if not all religions,[23] which could be (re-)activated as in the case of Lake Biwa. Moreover, many followers of the diverse religions do not respect freshwater and the environment to the extent suggested in theory (e.g., in the scriptures). This leads to the question whether and in what way religious beliefs play a relevant role in sustainable behaviour at all, and whether they may even hamper such behaviour. One might also wonder if religions can play both roles, destructive and constructive.

Can religion obstruct sustainable water management? Different perspectives from two case studies from Mali

In contrast to the above examples, it is indeed sometimes stated that religious beliefs and practices hinder sustainable development and water management. It has to be verified case by case whether this view is correct, or whether it reflects a tendency to suppress alternative views of local people.[24] Such a tendency may be motivated, for example, by disrespect for knowledge systems other than quantitative science and technocratic solutions, or by the characterisation of local people's religious viewpoints of environmental degradation and natural hazards as fatalistic. In particular, there is a danger that discrepancies between indigenous peoples' perceptions and "objective", science-based observations of changes in climatic, hydrologic and other environmental changes are used as a "proof" that religious interpretations are invalid.

In the following, I briefly summarise Helen Hintjens' analysis[25] of the different ways that indigenous belief systems were regarded in the context of two developmental projects in the Sahelian zone of Mali, funded by different non-governmental organisations. In one project – a tree seedling and selling activity in conjunction with diverse land and water management actions – attempts were made to "correct" perceptions and knowledge of the local Touareg people (characterised by a highly syncretistic Muslim/animist belief system), which were disrespected by the aid organisation's Christian and economic perspective. In particular, the fact that the natives consider trees to be a property of God which must not be sold was criticised – a view that is frequently expressed by Muslims and by indigenous peoples, also in debates about whether freshwater should be treated as an economic good.[26] The project leaders deemed that this somewhat fatalist attitude of the local people undermined their very chance for improving economic self-reliance (which was one goal of the sustainable development programme); and in the end, the concerns of the local people were not taken into account.[27]

By contrast, in the other project – a water management initiative which aimed to construct a well-digging scheme – the (spiritual) argumentations of the local people were not regarded as irrational but taken serious. One motivation for this was that wells dug into spots chosen on the mere basis of scientific/geophysical criteria are likely to be abandoned soon, which was why spots chosen according to the traditional (spiritual) criteria of the native pastoralists were given priority and finally turned out to be the more effective and sustainable solution.[28] This example highlights that the Western materialist standpoint of water and land management should not be forced upon indigenous people – rather, it may be much more effective to integrate their alternative, traditional knowledge forms and worldviews, including the religious aspects so dominant in indigenous communities.[29]

CLIMATE AND ENVIRONMENTAL CHANGE: THE FATE OF RELIGIOUS RITUALS

While the following case study represents another example of a mismatch between Western scientific and indigenous perspectives, it additionally points to the question of how traditional religious practices may be changed in response to ongoing and potential future climatic and hydrologic changes.

The study led by Joana Roque de Pinho[30] demonstrates that religious rituals and knowledge systems of the Maasai culture in the Kenyan Kajiado District are endangered by different socio-economic and environmental processes, including climate change. The Kajiado district is characterised by a highly variable and hardly predictable precipitation regime; hence, rain is so crucial to the pastoral livelihoods that "rain" and "God" (the female creator *Enkai*) are synonymous. The Maasai therefore have developed their own sophisticated weather forecasting system, which enables them to determine the beginning and the cessation of a rainy period by a suite of astronomical, meteorological, faunal, and other ecological indicators. In periods of actual or anticipated droughts, Maasai people perform rituals in the form of animal sacrifices and other ceremonies (such as walks over pastures, fertility songs and dances), in order to invoke the rain god to bring forth rain.

For many Maasai in this region, however, *Enkai* "has not come" anymore in recent years, that is, rain is perceived as becoming less frequent and more variable, which – among other factors such as formal education and an increasing Christian influence – tends to erode the traditional forecasting system[31] and, I presume, probably also the effectiveness and value of the rain-making rituals. The Maasai's observation of changing rainfall patterns, however, is not confirmed by instrumental meteorological observations. This mismatch has been explained by cognitive biases such as the over-weighing of recent drought events and the incomplete remembrance of historic events (which is a feature of most if not all humans). Furthermore, the Maasai may tend to attribute their recent problems of coping with dry conditions to ongoing climatic changes, while in fact these were primarily due to decreased mobility and due to higher vulnerability to droughts as a corollary of regional land use changes (for tourism, cultivation and wildlife conservation) and other socio-economic transformations.[32] However, the meteorological recordings to which the Maasai's perceptions were compared are rather uncertain, since they only crudely capture the spatial and temporal variability experienced by the nomads.

Whatever role climate change and related environmental impacts currently plays for the predicament of the Maasai: these changes remain a likely candidate to radically transform the Maasai knowledge and meaning systems – especially since a number of climate models project higher future temperatures and increasing rainfall variability in this region, at least during part of the year.[33]

Fatalistic attitude towards natural hazards? Floods and Allah's will

In her account of the strategies of Bangladeshis – of which 87% are Muslims – to cope with the frequent hazardous flood events in this region, Hannah Schmuck notices that most people regard these disasters as God-given, as Allah's will against which one cannot and should not do anything but pray.[34] Such an attitude is sometimes regarded as an inability of people to protect themselves and therefore as a factor that hampers adaptation to dangerous environmental change. At least, it appears in many cases that people's subjective perceptions of their means to cope with natural disasters do not match with the actual, objective possibilities.[35] Yet, this interpretation can be a misleading one: as Schmuck's Bangladesh case study demonstrates, people are not just passively accepting their fate but make the best out of their situation during the floods and ensure that daily life thereafter is secured (e.g., by fixing their beds below the roofs of their houses, or by switching to income resources other than crop production, etc.) – methods that have proven efficient for generations. Importantly, people always trust in Allah, which helps them to prevent feelings of being lost. Thus, as Hannah Schmuck concludes,[36] religion can serve as a healthy reaction and creative self-help strategy to overcome environmental hazards and return to daily life.

Pure or impure? Hinduism and India's polluted rivers

Hundreds of Hindu texts describe water as a sacred manifestation of goddesses with purifying qualities. Bathing in the rivers, drinking from them, or worshipping gods and goddesses in one of the many other water bodies and sacred sites along the rivers (as done by millions of pilgrims), brings salvation and spiritual awakening. For example, the tributaries of the Ganges river ("Mother Ganga") are perceived as absorbers of people's "moral dirt", which are purified themselves when they combine with the main river.[37] The conception of purifying rivers seems to contrast with the observation that many rivers in India – including the Narmada, the Ganges, and its tributaries – are heavily polluted through industrial and agricultural effluences.[38] As described in a recent issue of *The Economist*,[39] "Hinduism's sacred river contains 60,000 faecal coliform bacteria per 100 millilitres, 120 times more than is considered safe for bathing. Four miles downstream, with inputs from 24 gushing sewers and 60,000 pilgrim-bathers, the concentration is 3,000 times over the safety limit. In places, the Ganges becomes black and septic. Corpses, of semi-cremated adults or enshrouded babies, drift slowly by." Also, during the famous Maha Kumba Mela festival, more than 30 millions of people bathe in the river and also "scatter the ashes of their loved ones into the waters to mingle with the purity and divine wisdom symbolized in the rivers".[40]

In spite of the obvious consequences not only for the river ecosystems themselves but also for human health when using the thus contaminated water for daily life or for religious rituals, there is almost no official protest by the believers, including the religious leaders, against this dreadful situation. This is particularly surprising given the fact that the Vedas repeatedly refer to the importance of clean water, and that they forbid the pollution of water and of the environment in general.[41] This suggests "a separation of personal piety from the fate of the natural world as impacted by the devotee's own behaviour"[42] and from responsible action of Hindus towards protection of the environment (which is also found in other religions as well). As suggested by Vasudha Narayanan,[43] part of this contrast is due to the fact that the "Western" notion of physical dirt cannot pollute rivers thanks to their inherent purity – that is, a person's "moral dirt" can still be washed away even by heavily polluted waters. Narayanan offers additional explanations by suggesting that the goddess of wealth is regarded as more important than the goddess of Earth, and that Hinduism implies an emphasis on personal well-being and transcendence rather than an altruistic worldview in which water is to be kept clean for the benefit of all people. Therefore, it appears unlikely that religious motifs will stimulate policies for reducing the pollution of India's rivers and also for reducing e.g. the widespread unsustainable use of groundwater resources.

But, there are indications that this situation is beginning to change, and that the sacred texts and rituals help to mobilise people towards more environmentally sustainable behaviour.[44] Ironically perhaps, changes in attitudes and practices may follow in due course: rivers are found to be so heavily contaminated now that some religious leaders refused to participate in the rituals and started thinking about terminating the riverine rituals if river restoration measures are not taken soon.[45] And the World Bank has agreed to lend India one billion dollars over the next five years to clean up the Ganges.[46]

Perspectives

My present review of religious influences upon, and responses to, adverse (climate-driven) changes in water systems is certainly selective and by no means complete; nevertheless, it gives a glimpse of the diversity of functions that religion can have in this context. The starting point of this analysis was that religion – especially religions that regard nature and water as sacred and therefore as worthy of protection – may serve as a catalyst for sustainable behaviour (e.g. for mitigation of climate change or for avoidance of overuse and contamination of freshwater resources). Indeed, many religions show a high potential for motivating people towards environmentally friendly behaviour.[47] This has also been expressed more formally: for example, already the Assisi Declarations of 1986 stated that all religions should engage in "safeguarding the planet as a common inheritance", and also that no single religion should be blamed for the environmental crises.[48]

Since religious thoughts and feelings can increase altruistic behaviour even among anonymous people,[49] there is some hope that they could help recognising and eventually mitigating the climate change problem in particular, which is of a global nature and demands some form of global thinking and altruism. However, religion is certainly not the only powerful mechanism that determines one's environmentally relevant behaviour (especially in secularised societies):[50] it is interrelated with other cultural aspects; individuals may have their very personal religious reference system; and individuals frequently do not act in line with what the scriptures or other religious texts say. Therefore, *general* conclusions about the potential beneficial impact of specific (world) religions can as of yet hardly be drawn – this requires further field studies in different religious and cultural contexts. Also, this impact should not be overestimated and idealised.

In fact, as some other case studies have shown (see above), the role of religion in avoiding, and adapting to, dangerous climate and environmental change is not necessarily a favourable one but can be a neutral, an ambivalent, or an unconstructive one. An example for the ambivalent role of religion is the above-described sheer trust in Allah that characterises people in Bangladesh during flood events: this attitude can be regarded both as beneficial (because it helps the people to avoid the feeling of being totally lost and to go back to their normal life) and at the same time as fatalist (because it hampers more effective and precautionary adaptation techniques). An example of how religious practices can negatively influence the environmental status is the pollution of the Ganges and other rivers in India, which is partly due to the fact that masses of pilgrims bathe in the river and that sewages (including cadavers) from other religious rituals are thrown into the water. Further such examples are known from environmental history – at least, it is doubtful whether e.g. ancient (pagan and Christian) societies in Europe have cared better for their environment than today's secularised societies.[51] Also, in societies where the earth and the sky are considered separate domains and where God(s) are thought to control weather, the public understanding of the causes and implications of anthropogenic global climate change is limited, because a human influence on the climate is considered impossible.[52]

Most academic studies of religious aspects of climatic and hydrologic changes seem to be focused on indigenous communities. This may have several reasons. First, it is indigenous peoples whose traditional livelihoods, religious rituals and social cohesion are most severely affected by significant climatic changes and even more so by negative effects of prestigious water management projects such as the construction of large dams, irrigation canals, and the diversion of rivers, on which I do not focus here.[53]

Second, indigenous communities – especially those living in remote regions and/or regions characterised by variable or extreme climatic and environmental conditions – appear to be particularly vulnerable to climatic, hydrologic and other environmental changes. But obviously, the adaptive capacity of indigenous (and other, including industrialised) societies depends on whether or not these changes exceed the (natural) fluctuations to which people have learnt to adapt in the past. For example, while Bangladeshis are more or less successfully adapted to the frequent floods in their home areas – as shown by Hannah Schmuck's above-cited study – yet higher or more frequent floods in the future through more intense storms and sea level rise may require novel adaptation strategies.[54] However, one should also note that natural disasters are often regarded as a good reason for changing an individual's or a community's environmentally relevant behaviour, and that social cohesion and altruism can be triggered by, or intensified through, catastrophic events, e.g. through shared religious feelings.[55] Pascal Boyer even suggests that religious rituals can be regarded as "highly stylised versions of precautionary procedures", and deities as "highly simplified versions of absent human agents".[56] Nonetheless, the future is likely to hold climatic and environmental changes whose magnitudes and damages will exceed experiences from the past – and these new situations might eventually lead to rash, irrational and destructive actions,[57] irrespective of whether or not these involve a religious motivation. Capacity building through effective utilisation of "social capital" (social networks, solidarity) will therefore be crucial in disaster management.[58]

A third reason why indigenous people appear to receive a comparably large amount of attention in studies of religious aspects of climate and environmental change is that in the huge variety of their cultures and cosmologies[59] the environment is considered an integral part of the sacral cosmologies and daily rituals. For example, nature is perceived as animated, i.e., populated by land and water spirits who should be respected and honoured.[60] This fact suggests that the (religious) worldviews of these communities will be particularly affected by unusual environmental changes such as increasing weather changeability and resulting increases in the variability of hydrological processes: such changes always have a symbolic meaning and may distort, if not shatter, the foundations of these worldviews, like, for instance, rituals connected to the annual meteorological and agricultural cycle,[61] or traditional seasonal forecasting systems – compare the Maasai example discussed above. A similar case is reported for the Viliui Sakha people in the northeast of Siberia, where the recent gradual change in climatic conditions (wetter summers and milder winters, meaning the disappearance of a white bull, a personification of winter) is leading to a gradual if not abrupt transformation of the indigenous subsistence culture and their symbolic culture.[62]

A fourth reason might be that anthropology operates at the frontline of cultural studies and that anthropologists have focussed on non-Western cultures in their discipline. Therefore we still lack balanced research on different contemporary cultures worldwide. Exciting applications of approaches to study foreign cultures in studies of so-called secular cultures indicate that also the latter operate with cultural elements such as rituals, ultimate sacred postulates and not necessarily with rational values.

As the Maasai example also demonstrates, there can be significant disagreement between Western and indigenous knowledge about past, present and anticipated future changes in climate and environmental conditions (though there are case studies from other regions of the world which demonstrate that indigenous knowledge can also verify scientific measures).[63] Since one often cannot determine objectively which form of knowledge gives the "correct" description, it ought to be a fruitful approach to combine different perspectives. Ultimately, *respecting* the perceptions, needs and local theologies of indigenous communities and acknowledging them as perspectives as valuable as Western ideas can help to strengthen the adaptation capacity of local people.[64] Hence, traditional indigenous techniques of, for example, rainwater harvesting increasingly come to the fore and are being revived in many regions in the context of the new paradigm in water resources management. In effect, such methods appear to have a high potential for mitigating global problems of water scarcity and hunger.[65]

Another interesting feature of the new water management paradigm and the underlying revised water ethos is that it appreciates "holistic" concepts, in which human and environmental processes are regarded as closely interacting domains of complex socio-ecological systems, and in which water bodies and streams are regarded as multifunctional bloodstreams of all life in a watershed.[66] The new water ethos can be characterised by the recognition of freshwater as a vital resource and a common good that must not be denied to anyone; the involvement and empowerment of local people, regardless of sex and social position, in decisions about water use; the fair distribution and wise use of water within river basins; and a transparent flow of information about water availability, use, and decision making.[67] This paradigm shift reflects a major shift away from a utilitarian view toward an equity-oriented and more intrinsic view (which acknowledges that there is an inherent value to nature and water), and it could be aligned with (radical) theistic arguments that say that nature is a divine creation and should therefore be looked after by humans.[68]

THE NEED FOR STUDYING RELIGION IN CLIMATE CHANGE

The hydrological and other biophysical impacts of climate and environmental change in different corners of the world will have profound cultural and symbolic impacts, for example in that familiar places, lifestyles, identities, and religious rituals may get irretrievably lost.[69] Despite the obvious need for extensive and comparative ethnological and anthropological studies across a broad spectrum of regions and cultural settings, mental and spiritual implications of climate and environmental changes are, unfortunately, still systematically underexplored in climate change impact and adaptation studies (possibly because these impacts can hardly, if at all, be expressed in quantitative terms).[70] Such a systematic screening of religion's vulnerabilities and responses to climate and environmental change eventually ought to be embedded in a wider examination of "what is known about how human cultures (especially their ethics) evolve, and what kinds of changes might permit transition to an ecologically sustainable, peaceful, and equitable global society", as proposed by Paul Ehrlich and Donald Kennedy.[71] In other words, these authors plea for a sort of transdisciplinary "Millennium Assessment of Human Behaviour" to complement the existing Millennium Assessment of the world's ecosystem statuses.[72] This assessment will have to include studies of the circumstances under which people form new alliances that promote more sustainable ways of life – in fact, religious associations already play a significant role in this.[73]

At the same time that global climate and environmental change exposes specific religious rituals and symbols to the risk of loss or degeneration, religion and spirituality of all kinds could substantially contribute to the solution of environmental and water problems. For example, religious scriptures and myths could be mobilised for promoting environmentally friendlier lifestyles. It is even possible that without a spiritual and emotional dimension, humanity as a whole may never fully realise, and eventually solve, the massive and global-scale environmental problems including the climate change problem.[74]

One may argue from the perspective of enlightenment that a revitalisation of spiritual/religious viewpoints and myths is "irrational" and thus bears unforeseeable dangers. My personal reply to this argument would be threefold: first, religious viewpoints should of course not be the sole basis for decisions about the sustainable development of a region, but they should be critically reflected and carefully taken into account. Second, the environmental problems of the world are so enormous that any alternatives to the prevailing technocratic-economic path need to be explored, if they offer a chance to lead out of the current predicament (while ensuring that they can be abandoned if they prove to be misleading). Religious worldviews appear to be promising in this regard, as they provide fundamentally different views of the role of humans in this world. And finally, I would argue that it is also a myth – and a dangerous one at that, as environmental history has shown – that the world's natural resources, including freshwater, are infinite and therefore can be uncaringly exploited.

NOTES

1 Nelson, 70.
2 Parry et al.
3 Turner, MEA, Haberl et al., 12942-12947.
4 Postel, Bates et al.
5 Crutzen, 23.
6 E.g., Gosling et al., 299-341, ACIA, 649-944.
7 Perch-Nielsen et al., 375-393.
8 Diamond, Adger et al.
9 Stern, 526.
10 Magistro and Roncoli, 92; Leduc.
11 Klein et al., 750.
12 Grothmann and Patt, 200.
13 In this text, I use "religion" mostly for "visible", or "explicit", religion, while "invisible" or "implicit" religion includes in a more broader sense spirituality, attitudes, values, worldviews, visions, etc.; see Bergmann and Gerten (introduction to this book).
14 E.g., Gerten (2010).
15 Bates et al.
16 Pahl-Wostl et al., 484, Gerten (2008), 399-400.
17 Chamberlain, 3. Gardner suggests that religious worldviews inform 83% of the world population.
18 E.g. Chamberlain, 171.
19 <http://www.indigenouswater.org>, accessed 22 January 2009.
20 UNESCO/COMEST, 17-24.
21 Ibid., 20.
22 UNESCO/COMEST, 21.
23 Taylor, 991.
24 This view sometimes surfaces even in governmental as well as non-governmental programs for sustainable development, possibly because "religious beliefs may contain critiques of power structures in society, and moral misgivings concerning the misuse of natural resources" (Hintjens, 212).
25 Hintjens, 211-230.
26 Al-Jayyousi, 33-38; see also Northcott.
27 As a side note, animists sometimes ascribe the erosion of traditional methods of resource use to the influence of monotheistic religions, namely Islam and Christianity. In the present case, an elder's opinion is that the local people's respect for nature began to decline when these religions arrived in their area; see Hintjens, 224.
28 Ibid., 219.
29 Huntington, 1270-1274, Yohe et al., 833.

30 De Pinho et al., 1.
31 Ibid.
32 Western and Nightingale, 1, de Pinho et al., 1.
33 Bates et al., 782-788.
34 Schmuck, 88.
35 See Grothmann and Patt, 202-203.
36 Schmuck, 91-92.
37 Narayanan, 192.
38 Chamberlain, 17-18.
39 The Economist.
40 Chamberlain, 19.
41 Dwivedi, 10-12.
42 Ibid., 19.
43 Narayanan, 198.
44 Chamberlain, 20, 159, and references there.
45 Kürschner-Pelkmann.
46 BBC News, 3 December 2009, <http://news.bbc.co.uk/2/hi/8392118.stm>.
47 Taylor, 991-1008.
48 See <http://www.nyo.unep.org/eafi2.htm> for excerpts. Note that religious and other societal values have been found to be of little importance for the cause, degree and kind of environmental changes (see Turner II and McCandless, 230). This, however, does not preclude the possibility that religious beliefs can function as catalysts for more environmentally friendly behaviour.
49 Norenzayan and Shariff, 58-61.
50 E.g., Stern, 525-526.
51 E.g., Sonnabend, 82-89.
52 Donner, 233-234. Donner emphasises in this context that even in Western societies many people believe that natural disasters, like Hurrican Katrina in 2005, are deliberate acts of God.
53 Nelson, 81-82, Groenfeldt, 239.
54 This also holds true particularly for water management, as past experiences will not suffice to meet the future challenges under a changing climate; see Milly et al., 573.
55 E.g., Böhme, 8, Schmuck, 91. This phenomenon was also observed during the 2002 Elbe flood in East Germany, when many citizens helped the victims even if they were not affected themselves (but, religion was not necessarily a motivating factor for this prosociality).
56 Boyer, 1039.
57 Welzer, 276.
58 Mathbor, 360.
59 E.g., Solón, 36-38.
60 E.g., King et al., 390.
61 Wacker, 92.
62 Crate, 573.
63 E.g., Nichols et al., 68-79.
64 E.g., Fisher et al., 6, King et al., 285-390, and references therein, Rubow, 88-113.

[65] Rost et al.
[66] Falkenmark and Rockström, Gerten (2008); see also Nelson, 78-79.
[67] UNESCO/COMEST, 6-7.
[68] Armstrong, 11. While Clare Palmer argues against using the term "stewardship" because a steward's attitude still implies a separation of humans and nature and because this term suggests nature to be in need of human management (Palmer, 69-75), others argue from the point of view of evolutionary systems theory that the Earth of course needs human assistance for evolution toward a higher state (Kirchhoff, 70-77, 90-91).
[69] Adger et al., Turner et al.
[70] Bergmann.
[71] Ehrlich and Kennedy, 562.
[72] MEA.
[73] Ehrlich and Kennedy, 563.
[74] Groenfeldt, 237.

References

ACIA (2005), *Arctic Climate Impact Assessment – Scientific Report*, Cambridge: Cambridge University Press.

Adger, W.N., Dessai, S., Goulden, M., Hulme, M., Lorenzoni, I., Nelson, D.R., Næss, L.O., Wolf, J. and Wreford, A. (2009), "Are there social limits to adaptation to climate change?" *Climatic Change*, Vol. 93, 335-354.

Armstrong, A. (2006), "Ethical issues in water use and sustainability", *Area*, Vol. 38, 9-15.

Bates, B.C., Kundzewicz, Z. W., Wu, S. and Palutikof, J. P. (eds.) (2008), *Climate Change and Water: Technical Paper of the Intergovernmental Panel on Climate Change,* Geneva: IPCC.

Bergmann, S., "Dangerous environmental change and religion: How climate discourse changes the perception of our environment, the spiritual fabrication of its meaning and the interaction of science and religion", in this book.

Bergmann, S. and Gerten, D., "Religion in climate and environmental change: Towards a symphony of voices, memories and visions in a new polycentric field", in this book.

Böhme, H. (1988), "Umriß einer Kulturgeschichte des Wassers: Einleitung", in: Böhme, H. (ed.), *Kulturgeschichte des Wassers*, Frankfurt am Main: Suhrkamp, 7-42.

Boyer, P. (2008), "Religion: Bound to believe?" *Nature*, Vol. 455, 1038-1039.

Chamberlain, G.L. (2007), *Troubled Waters: Religion, Ethics, and the Global Water Crisis,* Lanham: Rowman & Littlefield.

Crate, S. A. (2008), "Gone the Bull of Winter? Grappling with the cultural implications of and anthropology's role(s) in global climate change", *Current Anthropology*, Vol. 49, 569-595.

Crutzen, P. J. (2003), "Geology of mankind", *Nature*, Vol. 415, 23.

Diamond, J. (2004), *Collapse: How Societies Choose to Fail or Succeed*, Viking Adult.

Donner, S.D. (2007), "Domain of the gods: An editorial essay", *Climatic Change*, Vol. 85, 231-236.

Dwivedi, K. D. (2003), "Ecological thoughts in the Vedas", in: Narayan, R. and Kumar, J. (eds.), *Religion and Ecology*, New Delhi: Deep & Deep Publications, 3-24.

The Economist (2008), "A special report on India: Creaking, groaning", *The Economist*, 13 Dec. 2008, <http://www.economist.com/specialreports/displaystory.cfm?story_id=12749787>, accessed 30 January 2009.

Al-Jayyousi, O. (2001), "Islamic water management and the Dublin Statement", in: Faruqi, N.I., Biswas, A.K. and Bino, M.J. (eds.), *Water Management in Islam*, New York: United Nations University Press, 33-38.

Ehrlich, P.R. and Kennedy, D. (2005), "Millennium Assessment of Human Behaviour", *Science*, Vol. 309, 562-563.

Falkenmark, M. and Rockström, J. (2004), *Balancing Water for Humans and Nature*, London: Earthscan.

Fisher, J.B., Nawaz, R., Fauzi, R., Nawaz, F., Sadek, E.S.S.Md., Latif, Z.A. and Blackett, M. (2008), "Balancing water, religion and tourism on Redang Island, Malaysia", *Environmental Research Letters*, Vol. 3, 024005, 1-6.

Gardner, G (2003), "Engaging Religion in the Quest for a Sustainable World", in: Starke, L. (ed.), *State of the World*, New York: W. W. Norton and Company, 152-175.

Gerten, D. (2008), "Climatic change, aquatic science, multiple shifts in paradigms", *International Review of Hydrobiology*, Vol. 93, 397-403.

— (2010), "Hydrolatry in Early European Religions and Christian Syncretism: How water transcends religions and epochs", in: Tvedt, T. and Oestigaard, T. (eds.), *Ideas of Water from Ancient Societies to the Modern World* (A History of Water Series II, Vol. 1), London: I.B. Tauris, 323-342.

Gosling, S.N., Lowe, J.A., McGregor, G.R., Pelling, M. and Malamud, B.D. (2009), "Associations between elevated atmospheric temperature and human mortality: A critical review of the literature", *Climatic Change*, Vol. 92, 299-341.

Groenfeldt, D. (2008), "Animism, economics and sustainable water development", in: Shaw, S. and Francis, A. (eds.), *Deep Blue: Critical Reflections on Nature, Religion and Water*, London: Equinox, 237-252.

Grothmann, T. and Patt, A. (2005), "Adaptive capacity and human cognition: The process of individual adaptation to climate change", *Global Environmental Change*, Vol. 15, 199-213.

Haberl, H., Erb, K. H., Krausmann, F., Gaube, V., Bondeau, A., Plutzar, C., Gingrich, S., Lucht, W. and Fischer-Kowalski, M. (2007), "Quantifying and mapping the human appropriation of net primary production in earth's terrestrial ecosystems", *Proceedings of the National Academy of Sciences of the United States of America*, Vol. 104, 12942-12947.

Hintjens, H. M. (1997), "What has God to do with sustainable development? A Sahelian dialogue", *Worldviews: Environment, Culture, Religion*, Vol. 1, 211-230.

Huntington, H.P. (2000), "Using traditional ecological knowledge in science: Methods and applications", *Ecological Applications*, Vol. 10, 1270-1274.

King, D.N.T., Skipper, A. and Tawhai, W.B. (2008), "Mori environmental knowledge of local weather and climate change in Aotearoa – New Zealand", *Climatic Change*, Vol. 90, 385-409.

Kirchhoff, J. (1998), *Was die Erde will*, Bergisch Gladbach: Lübbe.

Klein, R.J.T., Huq, S., Denton, F., Downing, T.E., Richels, R.G., Robinson, J.B. and Toth, F.L. (2007), "Inter-relationships between adaptation and mitigation", in: Parry, M.L., Canziani, O.F., Palutikof, J.P., van der Linden, P.J., and Hanson, C.E. (eds.): *Climate*

Change 2007: Impacts, Adaptation and Vulnerability, Contribution of Working Group II to the Fourth Assessment Report of the Intergovernmental Panel on Climate Change. Cambridge: Cambridge University Press, 745-777.

Kürschner-Pelkmann, F. (2008), "Ein Segen von oben", *Das Parlament*, Vol. 32, 9.

Leduc, T. (2009), "Climate research, interdisciplinarity and the spirit of multi-scalar thought", in this book.

Magistro, J. and Roncoli, C. (2001), "Anthropological perspectives and policy implications of climate change research", *Climate Research,* Vol. 19, 91-96.

Mathbor, G.M. (2007), "Enhancement of community preparedness for natural disasters: The role of social work in building social capital for sustainable disaster relief and management", *International Social Work*, Vol. 50, 357-369.

MEA (Millennium Ecosystem Assessment) (2005), *Ecosystems and Human Well-Being: Synthesis,* Washington D.C.: Island Press.

Milly, P.C.D., Betancourt, J., Falkenmark, M., Hirsch, R.M., Kundzewicz, Z.W., Lettenmaier, D.P. and Stouffer, R.J. (2008), "Stationarity is dead: Whither water management?", *Science*, Vol. 319, 513-514.

Narayanan, V. (2001), "Water, wood, and wisdom: Ecological perspectives from the Hindu traditions", *Daedalus*, Vol. 130, 4, 179-205.

Nelson, M.K. (2008), "Rivers of memory, lakes of survival: Indigenous water traditions and the Anishinaabe nation", in: Shaw, S. and Francis, A. (eds.), *Deep Blue: Critical Reflections on Nature, Religion and Water,* London: Equinox, 67-85.

Nichols, T., Berkes, F., Jolly, D., Snow, N. and the community of Sach Harbour (2004), "Climate change and sea ice: Local observations from the Canadian western Arctic", *Arctic,* Vol. 57, 68-79.

Norenzayan, A. and Shariff, A. F. (2008), "The origin and evolution of religious prosociality", *Science,* Vol. 322, 58-62.

Northcott, M. (2009), "Anthropogenic climate change and the truthfulness of trees", in this book.

Pahl-Wostl, C., Tàbara, D., Bouwen, R., Craps, M., Dewulf, A., Mostert, E., Ridder, D. and Taillieu, T. (2008), "The importance of social learning and culture for sustainable water management", *Ecological Economics*, Vol. 64, 484-495.

Palmer, C. (2006), "Stewardship: A case study in environmental ethics", in: Berry, R. J. (ed.), *Environmental Stewardship: Critical Perspectives – Past and Present*, London/New York: T & T Clark International, 63-75.

Parry, M.L., Canziani, O.F., Palutikof, J.P., van der Linden, P.J. and Hanson, C.E. (2007), *Climate Change 2007: Impacts, Adaptation and Vulnerability, Contribution of Working Group II to the Fourth Assessment Report of the Intergovernmental Panel on Climate Change,* Cambridge: Cambridge University Press.

Perch-Nielsen, S.L., Bättig, M. B. and Imboden, D. (2008), "Exploring the link between climate change and migration", *Climatic Change*, Vol. 91, 375-393.

Postel, S. (1997), *The Last Oasis: Facing Water Scarcity,* 2nd ed., New York: W. W. Norton & Company.

Roque de Pinho, J., Galvin, K. A., Boone, R. B., Zahran, S. and Bucini, G. (2008), "'God (rain) is not coming': Climate change and challenges to pastoralist weather forecast knowledge in Kenya Maasailand", poster presented at the Joint German-American Conference on Land Use and Climate Change, Berlin, Oct. 2-3, 2008.

Rost, S., Gerten, D., Hoff, H., Falkenmark, M., Rockström, J. and Lucht, W. (2009), "Potential to increase global crop production by rainfed water management", *Environmental Research Letters,* Vol. 4, 044002.

Rubow, C. (2009), "Metaphysical aspects of resilience: South Pacific responses to climate change", in: Hastrup, K. (ed.), *The Question of Resilience – Social Responses to Climate Change,* Det Kongelige Danske Videnskabernes Selskab (The Royal Danish Academy of Sciences and Letters), Copenhagen: Det Kongelige Danske Videnskabernes Selskab, 88-113.

Schmuck, H. (2000), "'An act of Allah': Religious explanations for floods in Bangladesh as survival strategy", *International Journal of Mass Emergencies and Disasters,* Vol. 18, 85-95.

Solón, P. (2006), "Cultural diversity and privatization of water", in: UNESCO (ed.), *Water and Indigenous Peoples* (Knowledges of Nature, Vol. 2), Paris: UNESCO, 36-39.

Sonnabend, H. (1999), *Naturkatastrophen in der Antike,* Stuttgart/Weimar: J. B. Metzler.

Stern, P.C. (2000), "Psychology and the science of human-environment interactions", *American Psychologist,* Vol. 55, 523-530.

Taylor, B. (2004), "A green future for religion?" *Futures,* Vol. 36, 991-1008.

Turner, B. L. II, and McCandless, S. R. (2004), "How humankind came to rival nature", in: Schellnhuber, H. J., Crutzen, P. J., Clark, W. C., Claussen, M., and Held, H. (eds.), *Earth System Analysis for Sustainability* (Dahlem Workshop Report 91), Cambridge and Berlin: The MIT Press and Freie Universität Berlin, 227-244.

Turner, B. L. II, Clark, W. C., Kates, R. W., Richards, J. F., Mathews, J. T., and Meyer, W. B. (eds.) (1990), *The Earth as Transformed by Human Action: Global and Regional Changes in the Biosphere over the Past 300 Years,* Cambridge: Cambridge University Press.

Turner, N.J., Gregory, R., Brooks, C., Failing, L. and Satterfield, T. (2008), "From invisibility to transparency: Identifying the implications", *Ecology and Society,* Vol. 13, 7, <http://www.ecologyandsociety.org/vol13/iss2/art7>.

UNESCO/COMEST (World Commission on the Ethics of Scientific Knowledge and Technology) (2004), *Best Ethical Practice in Water Use,* Paris: UNESCO.

Van Ufford, P.Q. and Schoffeleers, M. (eds.) (1988), *Religion and Development: Towards an Integrated Approach,* Amsterdam: Free University Press.

Wacker, C. (2006), "The flow of water and the continuity of culture: Water imagery in the landscape and rituals of a mountain desert oasis in Ladakh (Indian Himalaya)", in: UNESCO (ed.), *Water and Indigenous Peoples* (Knowledges of Nature, Vol. 2), Paris: UNESCO, 82-95.

Welzer, H. (2007), *Klimakriege: Wofür im 21. Jahrhundert getötet wird,* Bonn: Bundeszentrale für Politische Bildung.

Western, D. and Manzolillo Nightingale, D. L. (2005), "Environmental changes and the vulnerability of pastoralists to drought: A case study of the Maasai in Amboseli, Kenya", in: United Nations Environmental Programme (ed.), *Africa Environment Outlook Case Studies: Human Vulnerability to Environmental Change,* Nairobi: UNEP, 35-50.

Yohe, G.W., Lasco, R.D., Ahmad, Q.K., Arnell, N.W., Cohen, S.J., Hope, C., Janetos, A.C. and Perez, R.T. (2007), "Perspectives on climate change and sustainability", in: Parry, M.L., Canziani, O.F., Palutikof, J.P., van der Linden, P.J. and Hanson, C.E. (eds.),

Climate Change 2007: Impacts, Adaptation and Vulnerability, Contribution of Working Group II to the Fourth Assessment Report of the Intergovernmental Panel on Climate Change, Cambridge: Cambridge University Press, 811-841.

Life Breathing on Us

Three Painted Landscapes by Hardy Brix

Sigurd Bergmann

Hardy Brix, Arktisk Landskab (Arctic landscape), 2008, oil on canvas, 80 x 115 cm

Among all words in the preceeding and following chapters this chapter invites the reader to see, imagine and contemplate. Three painted landscapes, which are reproduced on the book's cover in all their complex colourfulness, form the center of the book and throw new light on our reasoning, dissecting and, I hope, also reconstructing of the theme. These images should not be misunderstood as simple illustration or as artistic comment from another angle; by contrast they a aim to bodily open our senses in order to approach and adopt environmental change in a deeper sense.

Hardy Brix, Nordisk Landskab 2 (Nordic landscape 2), 2006, oil on canvas, 38 x 50 cm

Hardy Brix' painted landscapes originate from a long encounter with arctic, subarctic and Nordic surroundings, which the artist has visited on research vessels and other boats over many years. The paintings emerge by standing on deck where the landscape can flow into the eye and further into the painter's "thinking hand" (J. Pallasmaa) covering the canvas with colours and shapes. The slow motion of the vessel carefully paving its way in this (for humans) strange environment can be grasped in them as well as the moving silence of a nature steadily in change and likewise continuously the same. Life-giving water flows through the landscape and the painting; the entanglement of land, water, light and form appeals to the observer's inner landscape, and it might evoke atmospheres and memories from other significant lands.

As the artist has visited these waters for many years, the paintings are by no means snapshots but compositions of an enduring deep encounter with a land- and seascape in continuous change. The reader might let them impact slowly and silently on his/her imagination before taking part of Petra Hansson's following chapter, which will connect the painted landscapes again to the wordy narratives of climate change.

The exercise to experience with the whole of one's bodily being not only painfully melting glaciers but the fragility of life in general on the one hand and the overwhelm-

Hardy Brix, Nordisk Landskab 4 (Nordic landscape 4), 2006, oil on canvas, 38 x 50 cm

ing grandeur of its incarnating power on the other, the experience to let life breath on us, might hopefully be encouraged by these painted environments. In my view such skill of perception is indispensable to every serious environmental scientist.

The "True" Story of Climate Change?

Aesthetic and Efferent Readings of Hardy Brix' Painted Landscapes

Petra Hansson

Introduction

We learn about climate change through a great variety of narratives. Climate change narratives have different purposes and contain different aspects of the complex array of intricate matters included in the discourse of climate change and they come in different forms such as texts, graphs, figures, pictures and photographs. A great variety of artistic exhibitions and initiatives explicitly addressing climate change have taken place lately which gives the impression that cultural responses and narratives to the challenges of climate change are increasing.[1] However, such responses to climate change have to a large extent been excluded from public debates on climate change.

This chapter contributes to the discussion of the arts within the climate change discourse and elaborates on what we can learn from artistic narratives in relation to technological and scientific climate change narratives. Considering education as one of the most important participants in the debate of sustainability and climate change, the burgeoning fields of Climate Change Education (CCE) and Climate Change Education Research (CCER) have a great responsibility in addressing questions of how we learn and understand climate change.[2] This chapter investigates literary critic Louise Rosenblatt's theory of "transactional reading" in relation to the Danish artist Hardy Brix' paintings of Arctic and Northern landscapes. Following Rosenblatt's transactional theory of reading and her ideas of "aesthetic" and "efferent" reading stances, two different interpretations of Hardy Brix' paintings are carried out. Even though Rosenblatt is mainly concerned with responses to and readings of literary texts, she is deeply concerned with aesthetic experiences in general. Therefore, I find her ideas useful in approaching visual images as well.

Climate change narratives

There is today a general consensus within the scientific community that anthropogenic climate change is a significant challenge to our culture. As the scientific evidence grows of the fact that the climate is changing and its effects on water systems, land use, and the difficulties connected to adapting and mitigating changes, communication of scientific evidence of environmental change to the general public becomes crucial in order to raise public awareness and increase knowledge of the issues involved. This means that understanding how climate change information is framed and narrated becomes central.

The extensive body of climate change science is one of the main sources serving the worldwide community with the latest facts of issues related to climate change. The Intergovernmental Panel of Climate Change (IPCC)[3] has an important role to play when it comes to "provide the world with a clear scientific view on the current state of climate change and its potential environmental and socio-economic consequences."[4] This "poetics of science" emphasises "an objective and complete assessment of current information".[5] In this sense, the IPCC becomes an influential participant in telling the "true story" of climate change.

Richard Hamblyn claims that climate change is often presented as "a persuasive narrative of risk and risk reduction."[6] He states that graphic presentations of scientific data have a pivotal role to play in communicating climate change to the public. He especially mentions the "Keeling Curve" describing the rise in carbon dioxide since 1958 which has been described as "an icon of global warming" and effectively used by Al Gore in the film *An Inconvenient Truth*.[7]

In her discussion of narratives in international climate policies such as UNFCC, the 1997 Kyoto protocol and the IPCC-reports, Diana M. Liverman mentions avoiding "dangerous climate change" as one of the key narratives.[8] Scientific facts are commonly illustrated with the help of climate models, graphic curves and global maps illustrating changes in temperatures. By using the colours red and orange to illustrate increasing temperatures in early climate science reports helps emphasising the danger functioning as "a striking image of a warming world."[9] Liverman concludes that the use of such images exclude human relationships to land from the narratives which makes the communication of "dangerous climate change" predominantly science based tending to leave human concerns such as vulnerability, adaptation and mitigation behind.

According to the latest IPCC report, the Polar regions (Arctic and Antarctic) are characterised as being "extremely vulnerable to current and projected climate change" and the regions also have "the greatest potential to affect global climate and thus human populations and biodiversity."[10]

From the "Arctic Climate Impact Assessment" (ACIA) of 2004[11] we learn that climate change is happening faster in the Arctic than anywhere else in the world. The average temperatures have doubled in the area compared to the rest of the world. Being especially vulnerable and at risk, the area is also considered to be the place the rest of the world look at when it comes to actual evidence of the fact that the climate is changing and the climactic changes taking place in the region will also ultimately "impact the planet as a whole."[12]

Michael T. Bravo claims that most climate change reports of the Arctic "portray Inuit and polar bears as the icons or victims of climate change."[13] He further states that indigenous "societies like the Inuit are used as exemplars because they epitomise the idea of being vulnerable, or 'at risk' " and regarding the Inuits, "vulnerability is expressed mainly in relation to sea ice."[14]

To sum up, dominating climate change narratives are "abstract, virtual and systemic"[15] dominated by scientific and technological explanations portraying a world at risk. As a result, we encounter natural scenes of mountains, icebergs, and oceans as diagrams and curves which give us the impression that climate change is something going on but nothing that concerns us.

Recent initiatives have emphasized the need of including local knowledge and artistic expressions to increase knowledge of climate change.[16] However, cultural expressions such as literature, art and music, often incorporated in local, indigenous and traditional knowledge systems, are seldom treated as a source of knowledge regarding climate change and its consequences in reports such as the ACIA. What then can artistic narratives contribute to the communication and understanding of climate change? And, how can they be interpreted?

Art and Transaction

In *Art as Experience*, John Dewey takes the Parthenon as an example claiming that it has "esthetic standing only as the work becomes an experience for a human being."[17] Dewey argues that there is a close connection between the fine arts and our everyday life and argues for "recovering the continuity of esthetic experience with normal processes of living"[18] emphasising the relationship between esthetic experience and happiness and regards these experiences as more valuable than just "compensating transient pleasurable exitations."[19] Consequently, Dewey claims that esthetic experiences can tell us something about the world. Transferring this to the discussion of what art can offer the climate change discourse, it seems plausible that artistic responses to climate change can tell us something about the matters involved.

In *Literature and Exploration,*[20] Rosenblatt develops a transactional theory of reading. She acknowledges the great potential and value of literature for enhancing the development of democratic citizens in a democratic society.[21] Rosenblatt emphasizes the importance of literature, and indeed all arts, in order to "prepare the student to meet unprecedented and unpredictable problems."[22] Accordingly, in relation to the uncertainties commonly connected to issues of climate change, I find Rosenblatt's theory useful in order to discuss what artistic responses can contribute to the climate change discourse. Rosenblatt argues: "In contrast to the analytic approach of the social sciences, the literary experience has immediacy and emotional persuasiveness."[23] Furthermore she claims that

> an object can have more than one value: it can yield the kind of fulfilment that we call *aesthetic* – it can be enjoyed in itself – and at the same time have a social origin and social effects.[24]

Another central idea in Rosenblatt's theory is the close relationship between education and the students' lives.[25] She argues that the object of study cannot be separated from students' previous experiences. Similar thoughts are put forward as being characteristic for indigenous knowledge systems emphasizing a relational approach to meaning making. Personal experiences are highly valued and considered crucial in order to make meaning of the world.[26] Further more, cultural expressions of the land are not separated from other ways of making meaning of the environment within indigenous knowledge systems. In Inuit traditions for example, songs and stories are regarded as having the ability to "establish and reinforce connections to the land and community"[27] and "they [direct] an individual's attention out into the land, to aid with his or her discovery of knowledge through personal experience."[28] Thus, cultural responses are regarded as more than mere representations of the surrounding environment.

Rosenblatt acknowledges the influence of Dewey for the move from "interaction" to "transaction".[29] Transaction differs from interaction in the sense that meaning is regarded as being created in the encounter between the reader and the text. This means that, the text becomes a text in the encounter between the reader and the text. Accordingly, interpretation is not static, but is rather regarded as a dynamic discussion between the interpreter (reader) and the object of study (the text).[30] Thus, taking a transactional approach to the reading of literature makes the idea of trying to understand the writer's ideas behind a text or trying to find the correct interpretation of a text impossible. Rather, the purpose of reading will always shape the meaning of the reading process.

According to Rosenblatt, our "readings" and understandings of literature are always individual and contextual at the same time and always related to the purpose of the reading and to the context of the reading processes. In this sense, different people *produce* literary works differently.[31] Therefore, my encounter and my analysis of Hardy Brix' painted landscapes can be regarded as an ongoing conversation between the paintings and me and in this sense also with the "actual" landscapes he is painting, i.e. the landscapes of the Arctic.

As mentioned above, Rosenblatt distinguishes between "efferent" and "aesthetic" stances. This line of thought is based on the idea that readers have different "focus of attention during the reading-event".[32] In an aesthetic reading, the attention is directed to what happens during the reading event which means that the reader pays attention to personal associations, feelings, and ideas.

> In aesthetic reading, the reader's attention is centred directly on what he is living through during his relationship with that particular text.[33]

The term "efferent" derives from the Latin "efferre" (to carry away) and implies that the purpose of an efferent reading of a text will generate some sort of information after the reading.

> As the reader responds to the printed words or symbols, his attention is directed outward, so to speak, towards concepts to be retained, ideas to be tested, actions to be performed after the reading.[34]

From this follows that the same text can be read either aesthetically or efferently depending on which attention the reader chooses, or which stance the reader adopts. However, Rosenblatt emphasizes that adopting an aesthetic stance is not just free associations, claiming that

> Perusal of a text merely leading to free fantasy would not be reading at all in the transactional sense. The concept of transaction emphasizes the relationship with, *and continuing awareness* of the text.[35]

Even though Rosenblatt distinguishes between the two stances, these two stances are not to be regarded as different poles; rather they are to be seen as a continuum the one or the other dominating the reading event.[36] Let us now turn to the main characters of this story, Hardy Brix and the Arctic landscapes.

Painting the Arctic

The first time I saw one of Hardy Brix' landscape paintings was at the transdisciplinary workshop *Religion and Dangerous Environmental-and-Climatic Change* in Trondheim in 2008.[37] As a complement to the more academic presentations at the workshop, Hardy talked about his boat travels around Greenland, Svalbard, Iceland and Antarctica at the same time as he showed us pictures of his paintings sprung from his travel experiences and encounters with such remote places. He did not explicitly touch upon the theme of climate change but his experiences of having been around the area had made him aware of issues related to climate change and his deep engagement in the area was evident.[38]

The beginning of the story I am telling here started during a walk in the streets of Trondheim. Hardy told me about his work and about his travels to the Arctic region and I told him about my fascination for the Arctic and my ongoing study of the significance of traveling in meaning making processes about nature in travel writer Jenny

Diski's works *Skating to Antarctica*[39] and *Daydreaming and Smoking about America with Interruptions*[40] in which she writes about her experiences of waterscapes and landscapes while on the move.[41] Brix seemed very interested in her work. He could clearly recognise Diski's fascination of "staring" at the sea from his journeys on cargo ships and we talked for a while about what it means to experience land and water without being "in" nature but by travelling through landscapes by train or by boat.

Brix' way of portraying the Arctic landscapes is clearly dependent on the way he is experiencing them from the ship's deck. He usually sits on deck capturing the light, the weather and the atmosphere painting quick snapshots of the landscape passing by. These paintings are small and it just takes a couple of minutes to finish them. This way of working resembles the idea of evoking an aesthetic experience from reading a text put forward by Rosenblatt.[42] In this sense, Brix is, in Rosenblatt's words, engaged in an "aesthetic drawing" or an "aesthetic painting process" while he is painting the landscape which he is encountering from his position on deck. His paintings of the Arctic landscape are results of his spontaneous and intuitive experiences of the landscape at the moment of painting. It is not that he has a schedule or a plan for what to paint every day. Rather, the paintings are produced in transaction with the present environment that he is encountering at the moment of painting. In the time of seeing.[43]

What happens when I approach these paintings taking an aesthetic stance? To me, at first glance, Hardy Brix' paintings appear blurred and vague. The colours are not very bright and variations of earthy colours dominate the pictures. His motifs are landscapes. Just landscapes. Uninhabited. The expected image of a dramatic Arctic emerges. A cold wintery landscape covered in ice. To me, the paintings also evoke a feeling of silence. Vast, endless and silent landscapes. There are no human beings or animals in the paintings. Just ice, mountains and the sky. I can hardly distinguish the mountains from sea from sky and it seems like land, ice, water and sky are one unit. No doubt the landscapes are grand. Dramatic. Full. Still there is movement in the paintings. I can easily imagine how huge the waves are and how cold it is. During his presentation in Trondheim, Hardy told us about the drops of rain or the snowflakes that sometimes land on his paper. The small paper covered by water mixing the colours. It is fascinating to think that in this sense, the landscape becomes a part of the painting. Bits and pieces of the Arctic landscape blend with the colours and move into the small paintings. It is fascinated to think about his paintings as bearing marks of the Arctic. In this sense, one specific moment of time and place is merged into the paintings.

Taking an efferent stance focusing on what information Hardy's paintings can give about climate change, the story becomes somewhat different. As stated previously, the Arctic is described as one of the most vulnerable areas on earth in terms of climate change. Climate change science narratives about the Arctic focus on ice melting and rising temperatures emphasising a place at risk. From this perspective, Brix portrays a threatened environment and in this sense the paintings tell us a story about a place, which is under constant and rapid change. However, I would not consider Brix' paintings as a portrait of an environment in stress or at risk. Even though his paintings are dramatic, there are no signs that tell us that he is portraying a particularly vulnerable area. Accordingly, the paintings tell us something more. It does not appear to be the Arctic as a "threatened environment" that Brix first pays attention to. It is rather the dramatic scenery, the colours and the landscape as a whole.

From my efferent study of Hardy's painting I have learned quite a lot of facts about the Arctic. Spending time analyzing the paintings took me out on new journeys to the Arctic through other kinds of images. For example, I have learned the geography of the Arctic region since I became curious to know where Hardy Brix had been. This led me to new investigations of other images of the Arctic as well. The most interesting observation is that many of the satellite images of the Arctic and Brix' paintings are remarkably similar depicting the Arctic in similar colours.[44]

Encountering the Arctic

The traditional western view of teaching and learning is based on the idea of knowledge as an object that can be transferred from one entity to another, from teacher to student or from texts to readers. Within this tradition, artefacts such as texts, graphs and pictures are often regarded as fixed objects that are supposed to be absorbed by the learner. Taking a transactional view of meaning making as starting point, knowledge and learning cannot be separated from personal experience. If knowledge and meaning are created in encounters through transactions between readers and texts it becomes impossible to separate the object of learning or study from the learner or the researcher. Adopting Rosenblatt's ideas means that aesthetic experiences such as art experiences play an important role in generating knowledge of our environment and thus of issues related to climate change. I believe Rosenblatt's way of acknowledging the close relationship between art and society has great potential in analyses of both pieces of art in themselves but also in analyses of the meaning of art in relation to understandings of climate change.

Taking a transactional starting point makes it possible to move beyond the gap between reality and representation, or the real and the simulated, and also to give up the idea of finding the truth. The idea that scientific observations of climate change give a more "authentic" and true description of issues related to climate change comes from a transactional perspective rather a question of the purpose of the "reading". Given the fact that the environment exists, it is also always mediated at the same time which makes it impossible for the real and the representation to exist without each other. From a transactional point of view, our encounters with our environments are always inflected by our previous experiences and they will affect our future encounters and the meaning of these encounters. Our experiences are contextualized and re-contextualized in every encounter. From a transactional perspective, it becomes unnecessary to pin out what is an authentic environment or a non-authentic environment and what is a direct encounter and what is a mediated encounter.

Communication of climate change science is traditionally based on empirical systematic knowledge derived from observation and experimentation by scientists such as ecologists, biologists and geographers. Using Rosenblatt's terminology, present day climate change narratives are dominated by an efferent "stance" having the purpose of educating the public about dangerous rises in temperatures and sea loss by using graphs, figures and maps. Regarding artistic stories as equally important starting points for discussions of climate change as scientific stories, I believe that cultural responses and readings of art can add unique knowledge and understanding of climate science that can do something else than tell us about the risks.

Is a piece of art an unobserved story of environmental phenomena such as climate change? For Hardy Brix' paintings, the answer is clearly no. His paintings are a part of his observations of the Arctic landscape and more than that, they could not have been produced without the encounter or the observation of the landscapes he is painting. In this sense, Hardy Brix' painted landscapes *are* landscapes. They are not based on scientific observation, but his encounters with the "actual" arctic landscape are what we as readers experience by encountering his paintings. Without Brix' "physical" encounter with the Arctic landscape his paintings would not exist.

In order to link aesthetic narratives to the climate change discourse and challenge the dominating paradigm of science narratives, we need theories for systematic analysis of aesthetic narratives in relation to environmental issues. In order for us to be able to judge and assess scientific information and critically discuss images and stories of climate change, our personal experiences becomes important. Following this, stressing Rosenblatt's idea of an additional aesthetic stance becomes crucial when it comes to the understanding of the complex issues connected to a changing climate.

The discussion of the meaning of reading for our experiences of the "natural" world and the relationship between reader and world is also discussed by ecocritic Scott Slovic. The fact that it is not possible to establish whether literature separates us or brings us closer to "reality" is put forward by Slovic:

> I now believe that books can either separate you from the world and from your life or deepen your engagement with reality. [...] Nature writing, then, can be a fence – an obstacle, an end in itself – or it can be a pathway to more intense awareness of ourselves and our environment.[45]

From a transactional point of view, Slovic addresses something important which can be related to Rosenblatt's discussion of aesthetic and efferent reading stances. If the aim of the reading has the efferent purpose of attracting attention to the physical environment, then the reading can be an eye opener, but if the aim of the reading is to direct the reader to a certain passage in the text in order to explain the plot or the language itself then it would probably not to the same extent help the reader to understand and perceive the environment in a "new" or "deeper" way and probably not enrich the reader's experience of the environment. The same would apply for art.

Our readings and our interpretations of landscapes are always shaped, formed and affected by our cultural, social and economical situations, our previous experiences. Consequently, we can have aesthetic experiences by both being in the Arctic and studying a piece of art of the Arctic. These experiences are not necessarily the same and one experience is not better than the other in terms of how we come to understand the Arctic. In this sense, the readers' experiences are crucial both for the understanding and perception of the environment and for the reading of art. And from an educational perspective it becomes important to highlight the two different stances which will generate different kinds of understandings. However, by including artistic narratives in the first place in the ongoing discussions of climate change also highlights what counts as relevant knowledge and who counts as a valid knowledge holder. Aesthetic experiences incorporate both physical and creative dimensions. From Hardy's paintings we can learn that cultural responses and artistic narratives give us the opportunity to explore both ourselves and the landscape. In this sense readings of artistic stories of the environment can help us to come closer to "vulnerable places" such as the Arctic and hopefully increase our understandings of them.

To me having dreamed about going to the Arctic since I was a child and always having Lisa Larsson's small statue of an Eskimo on my desk, Hardy's landscapes have taken me out on a journey to the Arctic. As close as I have ever been.

Notes

1 See for example <http://www.capefarewell.com/>, 9 September 2009, and <http://www.louisiana.dk/uk/Menu/Exhibitions/Green+Architecture+for+the+Future/>, 9 September 2009.

2 See for example <http://www.climatechangeeducation.org/>, <http://www.globalchange.gov/>, <http://www.unesco.org/en/esd/>, 7 September 2009. A forthcoming issue of "Southern African Journal of Environmental Education" is dealing with climate change education. See <http://www.eeasa.org.za/index.php?option=com_content&view=category&id=45&Itemid=72/>, 9 September 2009.

3 See <http://www.ipcc.ch/index.htm> for updates on previous, present and forthcoming reports published by the IPCC. It was in the 2007 report made clear that "It is very high confidence that the global average net effect of human activities since 1750 has been one of warming", (*Climate Change 2007: Synthesis report*, p. 36).

4 <http://www.ipcc.ch/organization/organization.htm/>, 9 September 2009.

5 <http://www.ipcc.ch/organization/organization.htm/>, 9 September 2009.

6 Hamblyn, 224.

7 Hamblyn, 232.

8 Liverman, 280. The other two key narratives are according to Liverman "the responsibility for climate change is common but differentiated", and "the market (in the form of carbon trading) is the best way to reduce the danger."

9 Liverman, 282.

10 Anisimov et al., 655.

11 Arctic Climate Impact Assessment (ACIA), <http://www.acia.uaf.edu/>, 9 September 2009.

12 ACIA, 8.

13 Bravo, 257.

14 Bravo, 262.

15 Daniels and Endfield, 215.

16 See for example Bates.

17 Dewey, 2.
18 Dewey, 9.
19 Dewey, 9.
20 Rosenblatt (1995).
21 Rosenblatt (1995), xv.
22 Rosenblatt (1995), 3.
23 Rosenblatt (1995), 7.
24 Rosenblatt (1995), 23.
25 Rosenblatt (1995), 3.
26 Bates.
27 Bates, 98.
28 Bates, 98.
29 Rosenblatt (2005), 40-41.
30 Rosenblatt (2005), 7.
31 Rosenblatt (2005),7-10.
32 Rosenblatt (1994), 23.
33 Rosenblatt (1994), 25.
34 Rosenblatt (1994), 24.
35 Rosenblatt (1994), 29.
36 Rosenblatt (2005), 12-14.
37 <http://www.hf.ntnu.no/relnateur/index.php?lenke=ridecc.php/>, 9 September 2009.
38 Hardy Brix’ paintings can be seen at his webpage <http://www.hardybrix.dk/>, 9 September 2009.
39 Diski (1997).
40 Diski (2002).
41 Cf. Hansson, Olsson Kronlid, Östman.
42 Rosenblatt (2005), chapter 3.
43 This way of working also parallels how nature writers such as David Henry Thoreau and Annie Dillard talk about nature writing in relation to the experiences of being in the landscape they are portraying.
44 Cf. *På randen af verden.*
45 Slovic, 180.

REFERENCES

Anisimov, O.A., D.G. Vaughan, T.V. Callaghan, C. Furgal, H. Marchant, T.D. Prowse, H. Vilhjálmsson and J.E. Walsh, *Polar Regions (Arctic and Antarctic). Climate Change 2007: Impacts, Adaptation and Vulnerability, Contribution of Working Group II to the Fourth Assessment Report of the Intergovernmental Panel on Climate Change*, ed. by M.L. Parry, O.F. Canziani, J.P. Palutikof, P.J van Der Linden and C.E. Hanson, Cambridge: Cambridge University Press 2007, 653-685.

Arctic Climate Impact Assessment (2004), *Impacts of a Warming Arctic*, Cambridge, UK: Cambridge University Press. Available at <http://www.acia.uaf.edu/>, accessed 9 September 2009.

Bates, Peter (2009), "Learning and Inuit knowledge in Nunavut, Canada", in: Chiba, M., Bates, P., Kube, S. and Nakashima, D. (eds), *Learning and Knowing in Indigenous Societies Today*, Paris: UNESCO.

Bravo, Michael T. (2009), "Voices from the Sea Ice: The Reception of Climate Impact Narratives", *Journal of Historical Geography*, Vol. 35, 2, 256-78.

Daniels, Stephen, and Endfield, Georgina H. (2009), "Narratives of Climate Change: Introduction", *Journal of Historical Geography*, Vol. 35, 2, 15-22.

Dewey, John (2005), *Art as Experience*, New York: Penguin books.

Diski, Jenny (2004), *Stranger on a Train*, London: Virago.

— (2005), *Skating to Antarctica*, London: Virago.

Garrard, Greg (2004), *Ecocriticism*, London/New York: Routledge.

Hamblyn, Richard (2009), "The Whistleblower and the Canary: Rhetorical Constructions of Climate Change", *Journal of Historical Geography*, Vol. 35, 223-236.

Hansson, Petra, Olsson Kronlid, David, Östman, Leif (in progress), *Encountering Nature on the Move: Meanings of Mobility in Jenny Diski's Travels.*

IPCC (2007) "Summary for Policymakers", in: *Climate Change 2007: Impacts, Adaptation and Vulnerability, Contribution of Working Group II to the Fourth Assessment Report of the Intergovernmental Panel on Climate Change*, ed. by M.L. Parry, O.F. Canziani, J.P. Palutikof, P.J. van der Linden and C.E. Hanson, Cambridge: Cambridge University Press, 7-22.

Liverman, Diana M. (2009), "Conventions of Climate Change: Constructions of Danger and the Dispossession of the Atmosphere", *Journal of Historical Geography*, Vol. 35, 2, 279-96.

På randen af verden, Exhibition catalogue ed. by Museet for religiøs kunst, Lemvig, 2009.

Rosenblatt, Louise M. (1994), *The Reader, the Text, the Poem: The Transactional Theory of the Literary Work*, Carbondale: Southern Illinois University Press.

— (1995), *Literature as Exploration*, 5. ed. New York: Modern Language Association of America.

— (2005), *Making Meaning with Texts: Selected Essays*, Portsmouth, NH: Heinemann.

Slovic, Scott (1992), *Seeking Awareness in American Nature Writing: Henry Thoreau, Annie Dillard, Edward Abbey, Wendell Berry, Barry Lopez*, Salt Lake City: University of Utah Press.

Sentient Landscapes, Vulnerability to Rapid Natural Change, and Social Responsibility

Thomas Heyd

The notion of religion, and the concepts, values and practices that go with this idea, is contested and under debate. Even so, it is evident that for many peoples, and likely across immeasurable time, the parts of the universe that should be treated with respect are widely dispersed over the natural landscape, including both its living and non-living parts. One way in which this perspective has been made sense of is through the attribution of *sentience* to land and its living and non-living constituents. In the following I discuss the relation of vulnerability to rapid natural change by exploring the manner in which, for cultures thoroughly anchored in the land, landscape conceived as sentient implies "social responsibility" for landscape.

I begin by briefly discussing the notion of vulnerability to natural phenomena, and possible ways of addressing it. Next I introduce the idea of sentient landscapes and the sense of responsibility toward landscape, often expressed among peoples who are deeply rooted in the land. I continue with a discussion of the basis of the respect that *underlies* this sense of responsibility, and its significance for the project of lowering our vulnerability to nature's agency. I conclude that it is imperative to transform our present ways of conceiving of natural phenomena.

Vulnerability and adaptive responses

While the thought of rapid natural change may bring to mind images such as eroded mountain slopes or deep arroyos cutting through alluvial valleys, such change can be brought about in a variety of ways. As Karen Baltgailis of the Yukon Conservation Society pointed out to me, Yukoners, for instance, have been struggling with rapid landscape change clearly caused by *human* activity, such as unsustainable mining and forestry practices. Generally, though, rapid landscape change is brought about by multiple, combined factors, some anthropogenic and some not. The prospect of rapid landscape changes (especially changes perceived to be of a catastrophic sort), whatever their causes, should lead responsible decision makers to take precautionary action (or remedial action, if the event has already taken place).

In the city of Valencia, Spain, where I spent my adolescent years, the river Turia flooded about a decade before my arrival. Even fifteen years after the event, people would point out the line on some of the buildings, way over my head, up to which the muddy flood waters had reached. This river had a long history of flooding the city, causing considerable deaths and damage to property over the centuries. So, after the terrible 1957 flood, the city government decided to build a concrete-encased river diversion on the outskirts of town, thus leaving the original riverbed entirely dry.[1]

Not everywhere can the resources be found to address catastrophic events with such grand and costly engineering solutions, or with sophisticated devices aided by contemporary technology as have been developed, for example, to save Venice (for a time) from the rising sea.[2] Nor are most events that bring about rapid landscape change, such as large-scale storms, droughts, storm surges on low-lying coastal zones, extremely hot summers, earthquakes, and tsunamis, amenable to straightforward managerial or engineering solutions. Such solutions, moreover, may be problematic in various ways. For one thing, they can lull populations into a false sense of security that can prove to be disastrous in cases of relatively "rare high magnitude events, during which the coping limits of the engineered structure might be breached", as Nick Brooks[3] points out.

When high impact, natural events, such as floods, happen, populations that do not generally expect such eventualities will be found unprepared. Furthermore, such solutions, suitable perhaps for "normal floods" but not for the kind that happens every 500 or every 1,000 years, may lead to planning decisions that can turn out to be dangerous. For example, they may lead municipalities to declare land in floodplains fit for construction. This is the case in the Vancouver area, since parts of this metropolis are located in the floodplain of the Fraser River, only protected by an aging system of dykes. In the light of these considerations I think that, as individuals and as societies, we need to address basic questions, such as whether some of those potentially catastrophic events actually are preventable (since human activities contribute to their occurrence). Given that such events may cause human beings great harm in their lives, livelihoods, and homes, we need to ask how we should think about, and what general approach we should have toward, the natural phenomena involved.[4]

Volcanic eruptions, earthquakes, and floods have always been potential threats to human well-being, but their effects tend to be limited to relatively circumscribed areas.[5] Lately though, global climate change has become an important topic of discussion, both for experts and for the general public. Foreseeably the phenomena that characterize climate change will be an issue for all human beings, since the warming of the globe is expected to bring about important, and relatively rapid, landscape changes in terms of desertification, thaw of permafrost in Arctic regions, reduction of the size and number of glaciers, rising sea levels, and so on, in many parts of the world. Reference to climate change may be useful to us in order to survey some of the issues that are generally relevant in the context of human responses to rapid landscape changes.

Climate change has brought into wide circulation the terms "prevention", "mitigation" and "adaptation". Even though it is a reasonable assumption – repeatedly supported by the Intergovernmental Panel on Climate Change (IPCC) – that human contributions to the current transformations of climate are very significant, it is now clear that the delay in the effects arising from its primary cause (the release of greenhouse gases) makes wholesale prevention of global climate change impossible by now. Given this conclusion, much recent policy discussion concerning human responses to climate change has been cast in terms of mitigation and adaptation. Internationally, and nationally at various levels (regional, provincial, city, etc.), mitigation has become a major concern, to be implemented through limitations on greenhouse gas emissions, and possibly through carbon sinks. Since it has become evident, moreover, that severe effects are going to be inevitable in many parts of the globe, adaptation increasingly is becoming an important complementary policy concern.

In the context of global climate change "vulnerability" is a key term. It can be understood in various ways. As Barry Smit[6] has noted, vulnerability is the product of distinct factors: exposure and sensitivity to exposure, on the one hand, and adaptive capacity or resilience to exposure, on the other. Neil Adger and Nick Brooks similarly propose that "vulnerability is not simply a function of exposure, but also of people's capacity to adapt to change. If the latter remains unchanged, increased exposure will lead to increased vulnerability."[7] In other words, given a certain driver of landscape change, such as global warming, that can generate powerful storms, earth slides, droughts, flooding, and so on, vulnerability is a function not only of the objective physical characteristics of landscape features, such as low-lying coastal areas, steep mountain terrain, and so on, but on how ready people are to respond in an adaptive way to those drivers.[8]

Researchers on climate change processes have been discussing a variety of possible physical adaptations to manage such phenomena. In areas with low-lying lands close to the sea, such as in the delta region of Bangladesh, building breakwaters and extended embankments in order to prevent flooding is being considered.[9] Another way to absorb change is in economic and social terms. People in the North African Sahel region, for instance, have responded to increased droughts by diversifying land use, moving from irrigated cash crops to more enduring subsistence crops, and by emigration to nearby cities in order to supplement incomes.[10]

Despite their obvious importance, I am not focusing on technological, socio-economic or managerial solutions. Directing our attention primarily to such ways of addressing problems posed by rapid natural change processes may well perpetuate a "fix-it" approach, which, ultimately, may be insufficient to overcome the challenges we face. A more fundamental approach may be needed, going deeper into the cultural fabric that animates all aspects of our interactions with the environment. There are some who address the need for adaptation by exploring the foundations of our cultural make-up. Thomas Homer-Dixon, for example, has proposed that in order to cope adequately with global climate change we primarily need to "front-load ingenuity".[11] As "business as usual" will not do, he proposes that, to generate the physical, social, and economic transformations needed to reduce vulnerability to climate change, we need to develop our sophisticated, human-specific, cognitive capacity to address and solve such problems.

Developing our capacity to generate ideas, and to apply them to practical situations, definitely constitutes an important ingredient in sharpening our coping capacities. But confronting adaptively the kind of situations that bring about rapid natural change also is a matter of acquiring appropriate ways of perceiving, and habits and practices that are suitable to the new situations at hand. The importance of ways of perceiving and of appropriate habits and practices can be illustrated in the context of outdoor activities. For instance, while cross-country skiing in mountainous terrain during spring weather one will sometimes face conditions that may result in an avalanche. Even though a highly developed ingenuity may be of use in such circumstances to devise strategies of avoidance or, in the worst case, tactics of remediation, such ingenuity may be insufficient if not accompanied by the ingrained precautionary habits and the relevant perceptual framework, such that circumstances are adequately assessed for their degree of danger.

Given that in contemporary urbanized societies people are largely divorced from the natural environment on which they depend, our ways of perceiving natural phenomena tend to be highly mediated. In order to grasp the possibilities of lowering vulnerability to rapid natural changes, I suggest that we take note of the ways of perceiving and valuing, and corresponding habits and practices, of peoples who have been long-time residents of places recognised as being subject to important landscape change drivers.

Sentient landscapes

While the exploration of cultural responses to natural phenomena, as manifested in various parts of the world, may reveal a variety of perspectives on the natural environment, this would take us too far afield. In this article I introduce just one account of responses to natural phenomena that bring about landscape changes, which illustrates a way of perceiving the biophysical environment that is very different from the ways of perceiving prevalent in our own, contemporary, Western societies.[12] The noted Canadian anthropologist Julie Cruikshank[13] writes about cultural responses to natural phenomena, in the northwest of North America, during a period of climate change prior to contact with Europeans, recounting some of the oral traditions about glaciers of the coastal Alaska Tlingit and the Yukon First Nations. She retells stories about glaciers that swallow up whole villages, but also points to the fact that glaciers served as a kind of "highway" that connected the interior of the continent with coastal areas. According to these oral traditions glaciers are not conceived of as inert, slowly sliding piles of ice but as entities that pay attention and respond to human behaviours, such as speaking carelessly, spilling blood, making noise, or cooking with grease in their vicinity.[14]

Cruikshank describes these peoples's way of conceiving the whole ensemble of entities, made up of human and non-human beings, including glaciers, by the term "sentient landscapes". This term takes note of the assumption that, from the Alaska Tlingit and Yukon First Nations' perspectives, the land is not just inert matter but alive, and capable of something akin to perception and action. To conceive of a stretch of land as a sentient landscape means that its diverse animate and inanimate components are not treated as mere resources (or mere obstacles, as the case may be) for human use, butas legitimate and, in some way, full *counterparts* to human beings.[15]

To people who have not been raised in the cultural milieux where these stories originate, the concept of sentient landscapes, and the accounts on which it is based, may seem incredible, even if, as Michael Chase points out, "the notion of the earth as animate is old and persistent, from Plato and Aristotle to Lovelock's Gaia hypothesis."[16] In any case, to focus on the divergence of the respective worldviews would be to miss the point. What is relevant in our context is that Cruikshank describes the type of relationship between people and land exhibited in these oral traditions as involving "social responsibility" arising from "the social nature of all relations between humans and non-humans, that is, animals and landscape features, including glaciers".[17] This approach to landscape, of course, is not unique to Alaska Tlingit and the Yukon First Nations but common to many peoples who have deep roots in their lands, including the Inuit and the Indigenous people of the Russian North as well as the Mapuche and Quechua, of South America's Andes mountain ranges.[18]

What is the normative import of this type of approach to land? Cruikshank points out that this "local knowledge embedded in oral traditions" displays "commitment to an active, thoroughly positioned human subject whose behaviour is understood to have consequences."[19] In her analysis, the type of relationship displayed in these approaches to landscape underscores "the social content of the world and the importance of taking personal and collective responsibility for changes in that world."[20] The basis of this sort of responsibility is worth exploring further.[21]

Self-organization and autonomy

In environmental ethics it is common to make a distinction between two sorts of duties or moral responsibilities. On the one hand, one may speak of responsibility *regarding* the natural environment, which stems from duties to other human beings who may be benefited or harmed by how we interact with the natural environment. On the other hand, one may speak of responsibility *to* natural entities themselves, on the supposition that those entities may also have a good of their own. Interestingly, the notion of "social responsibility" for landscapes described by Cruikshank seems not to fall squarely into either type of approach.

Generally, responsibility and duties generally are justified by appeal to the intrinsic (non-instrumental) value either of particular (usually human) entities. Social responsibility for landscape, as described by Cruikshank, crucially depends on the conceptualization of natural entities as *active and responsive*. As I explain more fully elsewhere,[22] responsibility to some entity minimally presupposes conceiving it as being structured in such a way that one can expect it to maintain its organization (at least for a time) in the presence of diverse forces. In this sense it is common in biomedical contexts to argue that the duty of medical staff to care for the well-being of their patients is a given, as long as the patients can continue functioning, at least at some level, but to accept that, in the case of brain death, there is a legitimate case for not continuing to keep patients "hooked" to machines since their capacity to function has ceased.

In order to clarify what sort of self-organization an entity has to have as a necessary condition for any responsibility to arise with regard to it, it seems appropriate to speak of *autonomy* in some sense.[23] The term "autonomy" may be apt here since it literally stands for being one's own law, or setting oneself one's own law, and, hence, implies the capacity for organising one's self,[24] even if all autonomy is relative, of course, since all beings are subject to some measure of influence or control by something beyond their own self.

Understood this way, to attribute autonomy to some entity means that it is not only organized in such a way as to maintain its unity and integrity (at least for a time) in the presence of a variety of forces, but that it may exert a systematic force on its environment, possibly actively but at least passively, through resistance to (actual or potential) influences. In other words, perceiving an entity as *autonomous* is perceiving it as *capable of maintaining its integrity*, and *of being sufficiently unified and dynamically structured to be both source and target of systematically effective forces*.[25]

According to this description of autonomy we certainly should grant that animals and plants are autonomous.[26] Animals seek to perdure, defending themselves, as far as they can, against aggression and sickness. Plants have ways of nourishing themselves, of countering pests, and even of controlling the effects of physical damage through mechanisms that seal cuts that would otherwise make them lose sap. The case for conceiving glaciers, tsunamis, and weather patterns, such as tropical storms, as autonomous is more problematic. Nonetheless, insofar as these entities and phenomena have systematic ways of affecting their environments, and their power is due to their specific kinds of integrity (evident by the difficulty in splitting these phenomena into their parts), they fit the pattern: All of these entities or processes may be organized sufficiently to maintain a certain unity over some time, and to show resistance to external forces. A snowflake that makes up part of a glacier, a drop of seawater moving up-shore, or a raindrop that constitutes part of a tropical storm, are all inoffensive when considered singly, but when constituting parts of particularly structured entities or phenomena, such as glaciers, flooding waters, or rainstorms, matters are otherwise. As such, glaciers, tsunamis, and tropical storms are best conceived of as "emergent phenomena".[27]

Consequently, the idea of social responsibility for our interactions with natural entities posited by Tlingit and Yukon First Nations, according to Cruikshank's account, makes a lot of sense if the entities in question are understood as having a certain capacity for autonomy, as here described. Social responsibility in relation to natural entities becomes most relevant in the present of "extreme" events, which may be disastrous for human beings. There presently is considerable research being carried out on the impact on behaviour and cultural perceptions following people's experience of disastrous natural events. Apparently, such events typically will remain salient in a society's cultural memory if the elapsed time span does not go beyond one lifetime. This seems to be confirmed among some populations living in Papua New Guinea and the Solomon Islands, for example, with regard to awareness of the signs of impending tsunamis and volcanic eruptions.[28] Such "extreme" event are seen as having a certain self-organization, which may lead to a feeling of respect for such phenomena, and consequent adaptive behaviours.

Such cultural perceptions may be oriented toward particular types of crisis situations, leading to particular coping behaviours, such as seeking high ground and ringing the alarm upon the recognition of the signs of impending events such as tsunamis. Other adaptive behaviours may be directed more toward the long term, such as the permanent relocation of villages or cities.[29] Sometimes the respect generated by the recognition of the autonomy of natural phenomena may lead to more indirectly adaptive behaviours, such as the creation of myths and the establishment of taboos about occupying certain areas of the land.[30] In those cases the direct cultural memory of the disastrous effects of the event may become lost but not before leading to an adaptation that exhibits respect for the natural phenomena at issue through habitual, ritual, or mythical means. So, although certainly not universally true, when people have been repeatedly exposed to phenomena that have sufficient self-organization to act in a unitary, possibly harmful way, one significant adaptation that people adopt seems to be the development of *respect* for these phenomena.

RESPECT

Respect is a complex notion, and it would take us too far afield to fully unpack it here. Suffice it to say that respect for natural phenomena may be of at least two sorts. On the one hand, people may feel compelled to respect some other being or process because of the perceived need to take care of themselves, as a mode of *self*-protection or precaution (if the effects that such phenomena and processes may have on their surroundings are taken to be significant). This is the sort of respect that we ordinarily speak of when we say that we need to "respect" the weather conditions when we travel in the high country or in Canada's winter, for instance. Similarly, people who have experienced a volcanic eruption or an earthquake may develop a respectful attitude toward volcanoes and areas near geological fault lines, respectively.

On the other hand, when the phenomena in question are conceived of as having such integrity and capacity to act that they mirror, even if only up to a point, our human capacity for maintaining integrity and for acting on the world, another form of respect also may seem relevant.[31] This kind of respect is less a matter of taking care of ourselves than of allowing these other phenomena sufficient space and time to express *their* self-organization. This is the sort of respect that we ordinarily accord fellow human beings who need quiet space and time to study or sleep, which leads us to take precautions so to avoid making noise, for example. More grandly, such respect is expressed in moral injunctions, such as to always treat humanity as an end and never as mere means.[32] This sort of respect may arise from a perception of common fates, which, in combination with a sense of community or conviviality, may lead to conscience and moral sense.[33]

"Social responsibility" for changes in the landscape, as described in Cruikshank's account, seems to arise from a combination of both of these kinds of respect. Insofar as natural phenomena can cause us trouble, we may want to take precautionary steps, and, insofar as the natural environment is seen as constituted by entities, which are self-organized enough to resemble us in relevant ways, we may want to establish something akin to social relation with them.[34] Though this may seem like anachronistic anthropomorphization of the natural world, today there are good reasons for treating natural entities as relevant counterparts to human beings.

Michel Serres, for example, argues that the situation of human beings in relation to the rest of nature calls for a new "natural contract", a contract analogous to a social contract among human beings.[35] This would be an agreement between human beings and the rest of nature such that the parties to the agreement can co-exist, and possibly even flourish each in its way, effectively requiring restrictions on the degree of human interference with nature. Such a contract would manifest a similar kind of respect for natural phenomena as is referred to by Cruikshank when she speaks of social responsibility for land.

In Serres's analysis, the consequences of our failure to agree on a "natural contract" in modern times have (mis)led human beings into activities that contaminate the natural environments with pollutants such that, ultimately, "natural services" of the sort taken for granted up to the present (clean water, clean air, productive land, and so on) increasingly are vanishing. A natural contract would seek to limit human activities in such a way that the flow of natural services may continue undiminished for present and future generations. In short, a natural contract would demonstrate something like social responsibility for the condition of the natural environment insofar as it expresses respect for phenomena that seem akin to us in their apparent capacity to act, to which we may be vulnerable.

Attention to vulnerability is especially relevant in situations in which natural forces may potentially transform landscapes in rapid ways that are catastrophic for human (and other living) beings. I propose that we may conceive of human vulnerability, in relation to phenomena that bring about rapid landscape change as a function of the attention given to the autonomy of significant natural entities and processes in our environment.

LOWERING VULNERABILITY

As noted, vulnerability depends not only on exposure to hazards but also on readiness in the face of drivers of potentially harmful change. The recognition of the relative autonomy of certain natural phenomena and processes, though, may lead to respect, which can contribute importantly to adaptive capacity. As explained, applied to the relation of human beings to their natural environment, respect may be conceived in at least two ways corresponding to the two ways discussed above: on the one hand, in terms of taking care of ourselves while, so to say, "in reach" or in the effective sway of those phenomena and processes, and, on the other hand, in terms of granting those natural phenomena sufficient "elbow room" (i.e., space and time) for their expression in our environment.

The idea of lowering human vulnerability by respecting natural phenomena and processes in this double sense has already been well understood by some environmental managers. Instead of trying to control rivers and the impact of the sea by raising levees or building more and higher sea walls, for example, some experts argue for the rehabilitation of deltas and polders as flood retention areas for rivers, and for the restoration of mangroves and coastal forests in the case of threatened coastal areas, respectively. Lowering vulnerability by taking note of the self-organization of natural phenomena in these kinds of ways may lead to the development of policies that lower vulnerability, both at the individual and societal level.

It is true, of course, that individuals may only have limited ability to move house or change the conditions in which they gain their livelihood. Nonetheless, as long as society makes certain material resources and know-how available, a lot of steps can be taken to adapt one's private space and one's workplace surroundings to potential environmental hazards. For example, in flood prone areas people have raised their houses or moved to upper stories.[36] In earthquake zones they reinforce buildings and firmly attach objects (such as bookshelves) to walls to prevent that they become hazards.

At the societal level, furthermore, respect for natural phenomena means not subjecting populations to unnecessary risks, and making the material, social and informational infrastructures available that can help individuals and communities address natural phenomena that may potentially be hazardous to them. Practically this may mean making it possible for people to relocate away from high to low risk areas, to take a proactive approach regarding landscape changes that are due to arrive in the course of time due to geological or geographical reasons (e.g., earthquakes in areas with fault lines, floods in floodplains) by facilitating appropriate building standards and by regulating the use of suitable building materials, and so on.[37]

Conclusion

Rapid environmental landscape change invites us to reflect on vulnerabilities and the ability to address natural phenomena appropriately. Confronting rapid natural changes effectively and responsibly certainly requires scientific research, in order to understand natural processes, and the application of our ingenuity to come up with appropriate physical and socio-economic modifications to our environment and our societies. Decreasing vulnerability and strengthening resilience, however, are, moreover, grounded in larger, more encompassing, cultural matrices.

Based on the analysis of the notion of social responsibility for changes in the world introduced by Cruikshank, I have suggested that vulnerability, in a more general sense, be understood, in part, as a function of the *conception of natural phenomena*, held by individuals and societal decision makers, and of the *values associate to those conceptions*. As Cruikshank notes, "our human ability to come to terms with global environmental problems will depend as much on human values as on scientific expertise."[38]

I conclude that, insofar as the recognition of autonomy of certain natural phenomena may lead to adaptive behaviour and policies, we urgently need to promote the conditions under which the recognition of such autonomy, and the acquisition of the corresponding deep respect and sense of responsibility for natural phenomena, may be achieved.[39] Even if we may not necessarily recognise such respect as a matter of religious belief or practice, it certainly is a way of approaching the natural world that has been the hallmark of religious expression throughout much of humanity's existence. This suggests that further exploration of such religious expression may be of great value to the enterprise of determining how we may build resilience to the risks arising from rapid natural change.[40]

NOTES

1 Gozalo de Andrés.

2 See Merali on the "Venice Tide Barrier Project" (Modulo Sperimentale Elettromeccanico, involving seventy-nine mobile floodgates.

3 Brooks.

4 I fully acknowledge that certain events generally classified as natural, such as storms and floods, may have a considerable anthropogenic component. Nonetheless, so long as the non-human natural contribution to the event is most significant, I will, for simplicity's sake, speak of natural events, phenomena, or processes. So, while I wouldn't count the radioactive contamination of the environment following an incident at a nuclear power station, such as happened at Kashiwazaki-Kariwa nuclear power plant 16 July 2007 leading to the spill of radioactive wastes into the sea, I do treat storms, such as Hurricane Katrina, which may partly be the result of anthropogenic greenhouse gases, as natural. (I realize that this is not a neat way of sorting out the world but, for further clarification, see Heyd (2007), ch. 9, "Nature, Culture, and Natural Heritage: Toward a Culture of Nature".)

5 But see, e.g., Burroughs on the capacity of "supervolcanoes" such as Toba, Hekla, or Thera, to alter climate worldwide and for considerable time spans.

6 Smit.

7 Adger and Brooks, 29.

8 Given the diversity of socio-economic and political situations in which people find themselves, vulnerability varies on both individual and collective levels. On how to think about vulnerability, also see Kelly and Adger.

9 Ahmed et al.

10 See Mortimore and Adams.

11 Homer-Dixon (2005). Also see Homer-Dixon (2000) and (2006).

12 There are other relevant accounts, of course. See, for example, Fagan, who contrasts the responses to natural phenomena, during climate change, of Sahelian herders, South Africa's San, and South America's Moche, among others.

13 Cruikshank (2001) and (2002).

14 Cruikshank (2001), 385, 387-388.

15 Also cp. Ingold's notion of "sentient ecology".

16 Chase.

17 Cruikshank (2001), 382.

18 When I write of having "roots" I do not intend to differentiate between nomadic and sedentary peoples, since nomadic peoples are also place-dependent. Also see Brody on the deep connection to particular stretches of land that hunter-gatherers have.

19 Cruikshank (2001), 391.

20 Cruikshank (2001), 391.

21 See Heyd (2007) for a fuller view on responsibility for the natural environment in which people are enmeshed, especially chapter 4, "Environment and Culture in Latin America: Community, Autonomy and Resistance".

22 See Heyd (2005).

23 Heyd (2005).

24 Also see Prigogine and Stengers on self-organizing systems, and Maturana and Varela on autopoeisis. I thank Michael Chase for pointing out to me the relevance in this context, moreover, of Kaufmann (1995) and (2000).

25 I thank Mark Woods for helping me clarify these points. He suggests that the autonomy of nature "contrasts with obedience: wild things are autonomous because they have not changed to adopt the imposed will of another. We can also think of autonomy in terms of authenticity: being self-expressing, self-actualizing, or self-realizing" (Woods).

26 But see Kant for giving a sense to the term "autonomy" that makes autonomy a property that exclusively may characterize human beings.

27 See Holland. I owe this reference to Michael Chase.

28 Davies, 37-38.

29 Davies, 39-40; also see Fagan on the Moche relocation of their capital.

30 Lowe et al., 138.

31 It is notable that there may be good adaptive reasons for our tendency to anthropomorphize diverse entities and processes in the natural environment in this way. See Burroughs. Phenomenologically it makes sense to attribute moral standing to those beings that resemble us because we can empathize and sympathize with them.

32 Kant.

33 Also see Heyd (2007), ch. 2.

34 It is notable that the feeling of respect and sense of responsibility may well arise with regard to phenomena and processes that one may not suppose capable of intentionality, understood as the capacity for making outright choices among possible courses of action.

35 Serres.

36 Climate Proofing.

37 For a list of possible steps that both individuals and society in general could take with regard to climate change, see Dauncey. On historic and prehistoric human responses to powerful environmental impacts that did, or could have, amounted to catastrophes for human societies, see, for example, Leroy and Diamond.

38 Cruikshank (2001), 390.

39 For further development of the topic of respect for, and appreciation of, landscapes and nature, see Heyd (2007).

40 I am grateful to Edward Butterworth, Tony Berger, and Jutta Gutberlet for attentively reading this chapter, and providing me with useful comments. I am also indebted to Nick Brooks and Michael Chase who have made a number of excellent suggestions for improvements of this version of this essay.

References

Adger, N. and Brooks, N. (2002), "Does global environmental change cause vulnerability to natural disaster?" in: Pelling, M. (ed.), *Natural Disaster and Development in a Globalizing World*, London: Routledge, 19-42.

Ahmed, Ahsan Uddin, Alam, M. and Atiq Rahman, A. (1999), "Adaptation to climate change in Bangladesh: Future outlook", in: Huq, S., Karim, Z., Asaduzzaman, M. and Mahtab, F. (eds.), *Vulnerability and Adaptation to Climate Change for Bangladesh*, Dordrecht: Kluwer, 125-43.

Brody, H. (2001), *The Other Side of Eden, Hunters, Farmers, and the Shaping of the World*, New York: North Point Press.

Brooks, Nick (2007), Personal correspondence, 24 July.

Burroughs, William James (2005), *Climate Change in Prehistory: The End of the Reign of Chaos*, New York: Cambridge University Press.

Chase, Michael (2007), Personal correspondence, 22 July.

Climate Proofing: A Risk-Based Approach to Adaptation (2005), Manila: Asian Development Bank.

Cruikshank, Julie (2001), "Glaciers and climate change: Perspectives from oral tradition", *Arctic*, Vol. 54, 4, December, 377-93.

Cruikshank, Julie (2002), "Nature and culture in the field: Two centuries of stories from Lituya Bay, Alaska", in: de Laet, M. (ed.), *Knowledge and Society (Research in Science and Technology Studies: Knowledge and Technology Transfer)*, Vol. 13, Amsterdam, JAI/Elsevier Science, 11-43.

Dauncey, Guy (2001), *Stormy Weather: 101 solutions to global climate change*, Gabriola Island, B.C., Canada: New Society Publishers.

Davies, Hugh (2002), in: Torrence, R. and Grattan, J. (eds.), *Natural Disasters and Cultural Change*, London/New York: Routledge, 28-32.

Diamond, Jared (2005), *Collapse: How Societies Choose to Fail or Succeed*, New York: Viking.

Fagan, Brian (2000), *Floods, Famine, and Emperors: El Niño and the Fate of Civilization*, New York: Basic Books.

Gozalo de Andrés, Carmen (2003), *Las Riadas del Turia en Valencia; parts I and II*, <http: //www.meteored.com/ram/numero16/riadaturia.asp> and <http://www.meteored.com/ram/numero16/riadaturia2.asp>, 18 July 2007.

Heyd, Thomas (ed.) (2005), *Recognizing the Autonomy of Nature: Theory and Practice*, New York: Columbia University Press.

— (2007), *Encountering Nature: Toward an Environmental Culture*, Aldershot: Ashgate.

Holland, John (1998), *Emergence: From chaos to order*, New York: Basic Books.

Homer-Dixon, Thomas F. (2000), *The Ingenuity Gap*, New York: Knopf.

— (2005), "Adaptive Capacity and Resilience: How to Meet Challenges from Climate Change", in: *Introduction to Climate Change: Impacts & Adaptation*, session C-CIARN (Canadian Climate Impacts and Adaptation Research Network) Conference, London, Ontario, 1 June 2005.

— (2006), *The upside of down: Catastrophe, creativity and the renewal of civilization*, Toronto: A.A. Knopf.

Ingold, Tim (1994), "Introduction to Culture", in: Ingold, T. (ed.), *Companion Encyclopedia of Anthropology, Humanity, Culture and Social Life*, London: Routledge.

Jickling, Bob (ed.) (1996), "What is a good way to teach children and young adults to respect the land? A panel discussion", transcript in: *A Colloquium on Environment, Ethics, and Education: Whitehorse, Yukon, Canada, July 14-16, 1995* (proceedings), Arts and Science Division, Yukon College, Whitehorse, Yukon, Canada, 32-48.

Kant, Immanuel (1993), *Grounding of the Metaphysics of Morals*, transl. by James W. Ellington, Indianapolis: Hackett.

Kaufmann, Stuart A. (1995), *At home in the universe: The search for the laws of the self-organization and complexity*, New York/Oxford: Oxford University Press.

— (2000), *Investigations*, New York/Oxford: Oxford University Press.

Leroy, Susanne (2006), "From natural hazard to environmental catastrophe: Past and present", *Quaternary International*, Vol. 158, Issue 1, December.

Lowe, D.J., Newnham, R.M. and McCraw, J.D. (2002), "Volcanism and early Maori society in New Zealand", in: Torrence, R. and Grattan, J. (eds.), *Natural Disasters and Cultural Change*, London/New York: Routledge, 126-61.

Maturana, Humberto and Varela, Francisco (1973/1980), in: Cohen, R. S. and Wartofsky, M. W. (eds.), *Autopoiesis and Cognition: The Realization of the Living*, (Boston Studies in the Philosophy of Science, Vol. 42), Dordecht: D. Reidel Publishing.

Merali, Zeeya (2002), "Saving Venice: An ambitious plan seeks to prevent a modern Atlantis", in: Focus, *Scientific American*, (August 19); <http://www.sciam.com/article.cfm?articleID=00088E1F-D709-1D5B90FB809EC5880000>, 18 July 2007.

Mortimore, M. and Adams, W.M. (2001), "Farmer adaptation, change and 'crisis' in the Sahel", *Global Environmental Change*, Vol. 11, 1, 49-57.

Prigogine, Ilya and Stengers, Isabelle (1984), *Order out of Chaos: Man's new dialogue with nature*, Toronto: Bantam.

Serres, Michel (1990), *Le Contrat naturel*, Paris: François Bourin, (*The Natural Contract*, Ann Arbor: University of Michigan Press 1995).

Smit, Barry (2005), "Where From and Where to: Climate Change Impacts and Adaptation Research and Practice", in: *Introduction to Climate Change: Impacts & Adaptation* session, C-CIARN (Canadian Climate Impacts and Adaptation Research Network) Conference, London, Ontario, 1 June 2005.

Sulyandziga, Pavel, and Vlassova, Tatiana (2001), "Impacts of climate change on the sustainable development of traditional lifestyles of the Indigenous People of the Russian North: Towards the development of an integrated scheme of assessment", *Northern Review*, No. 24 (Winter), 200-207.

Woods, Mark (2005), Personal communication.

What's in a Name?

Gaia and the Reality of Being Alive in a Relational World

Anne Primavesi

> Twenty years ago ... I had no clear idea of what Gaia was, nor had I thought deeply about her. What I did know was that the Earth was different from Mars and Venus. It was a planet with apparently the strange property of always keeping itself a fit and comfortable place for living things to inhabit. I had the idea that somehow this property was not an accident of its position but was a consequence of life on its surface ... The novelist William Golding thought that such an idea should be named Gaia after the Greek goddess of the Earth.[1]

James Lovelock's idea of Earth as a planet with the strange, indeed unique property of keeping itself a fit and comfortable place for living things to inhabit is now generally accepted. For over thirty years he and other scientists have been testing the idea in innumerable ways. Research has centred on the continuing contributions made to Gaia by living things themselves, from ocean algae to our own species. We now know that our contributions are a major factor in what we call "climate change" and all that that entails.

That being the case, we need to understand as fully as possible what it means to live within Gaian space and time in order to contribute to its wellbeing rather than its detriment. As it is the essential context for human life and evolution, we are not only totally dependent on its ability to keep Earth a fit and comfortable place for us to inhabit. We also contribute to its capacity or inability to maintain the lives of all other beings as well. Informing ourselves about the reality of being alive in the Gaian relational Earth system leads to the realization that we are not the dominant species within it but one kept alive by and therefore dependent on species other than our own. Our mutual interdependence on and within Gaia is the reason for our shared responsibility for the givenness and character of present and future life on Earth.

We need many different kinds of information about this state of affairs. Scientific, economic and, to a certain extent, philosophical knowledge about it is increasingly available to us. This is adding to our understanding of how we might live so as to sustain healthy relationships within the Gaian community of life. Christian teaching over the centuries has had a rather different emphasis. It has laid stress on the importance of present interrelationships for a future salvation in heaven for ourselves:

one that does not include all present and future life on Earth. Both implicitly and explicitly it has ignored and even promoted the violent character of our interactions with the more-than-human community and their effects on our shared environment. When made aware of it, all too often it has sanctioned an increase of that violence: in the name of an omnipotent, punitive God invoked on behalf of our self-interest alone. Exploitation of land and animal resources for financial gain has been formally and consistently blessed by Christian ministers throughout western Christendom and in its colonialist expansion across the world.

Now, however, we are becoming more and more informed about and aware of the effects of human violence on all members of the Gaian community. This is accompanied by a growing realization that we do not and cannot exist outside of it. The scientific information now available to us is also beginning to re-inform our religious understanding, leading it back into the unity that is already there between God and the whole community of life. As this re-information occurs, the scientific question about our unity within Gaia prompts a religious one: Who for us, who for me, is God?

Because whatever we may think or say on the theoretical level, in actual hard fact "God" is simply another word for "reality". My "God" is, in practice, that which my whole life tends towards, that which for me is most real.[2]

We now know that our real, earthly lives are shaped bodily, mentally and spiritually by the reciprocal relationships and interdependent character of our existence within Gaia. Therefore whether or not my life as an individual tends towards violence or nonviolence both contributes to and shapes the character of the whole, for good or ill. And as an individual, the "character" of my God will both influence, re-inform and re-inforce in practice that violence or nonviolence towards which my whole life tends.

This religious truth is explored in detail in the second part of this essay in a consideration of Bonhoeffer's response to his own question: "Who really is Jesus Christ for us today?" First, however, the scientific truth about the reality of Gaia will be presented as an argument for a nonviolent lifestyle in the climatic situation in which we now find ourselves.

Nonviolence informed by Gaian science

Acknowledging that we depend on other species of living beings to keep us alive calls for a major shift in our western thinking about ourselves. Industrialization and urbanization have, to a large extent, removed us from direct contact with those more-than-human lives that contribute to our essential nourishment. Overall, there is little active awareness of the all-pervasive interdependence that keeps us alive. Coupled with this is a resistance to acknowledging that through our present role in accelerating climate change we are endangering Gaia's ability to sustain the lives of those other species – and with them, our own.

Up to now, if we thought about it at all, we assumed that Gaia as a global system would continue to keep Earth habitable for human life during the whole of the planet's existence. That remains the case for most of us, even though the fact of our common dependence on the sun has led scientists to calculate the years remaining before the sun expands to engulf the Earth's orbit, and burns it up. Now however, as the potentially destructive nature of our own negative contributions to Gaia (such as excess carbon dioxide in the atmosphere) becomes increasingly clear, we are also becoming aware of what *on earth* makes our lives possible. It is the continuous tight bonding between all life and its physical and chemical environment.[3]

With some exceptions – such as the lifeforms who inhabit the ecosystems in and around the deep sea hydrothermal vents needing neither light nor oxygen to thrive – the nutritional basis of that complex web on the planet's surface is the sun-energized process of photosynthesis.[4] It evolved in microbes as soon as life emerged on the planet and, since then, at every level from microbe to whale, organic beings use air, water and other organic beings to build their reproducing selves. Local ecology has become global ecology and, as a corollary, life does not exist *on* Earth's surface so much as it*is* Earth's surface. Organisms are less self-enclosed isolated individuals than communities of bodies exchanging matter, energy and information with others:

> Each breath connects us to the rest of the biosphere, which also "breathes", albeit at a slower pace.[5]

To exclude ourselves, even notionally, from membership of the Gaian community or biosphere would effectively exclude us from life. For we exist only while Gaia exists. What keeps us alive is available to us only in so far as we share it with other living beings and they share it with us. The relational, total-field image of ourselves offered by Gaia is one in which the intrinsic relations between us belong to the definition of what we are. Without those interactive relations, I no longer exist as a person. On the canvas of life I am, so to speak, a knot in a field of intricate relations.[6] So a set-back for Gaia, such as an above average temperature rise, is a general set-back for the community of life and an individual set-back for me. Similarly a lowering of global temperature is both a general and an individual gain to which I as an individual may or may not contribute.[7]

The character of our Gaian relationships

This last comment by Arne Næss refers to the fact that each of us may choose, or not, to contribute to the welfare of the whole by, for instance, reducing our carbon emissions. Our ethical autonomy determines the freedom of our actions within Gaia while our dependence on each other for nourishment and our common sharing the planetary resources of air, earth, solar energy and water focusses attention on the way in which we use that freedom. Do we, immediately or by default, use it to do violence to others? That is, do we, literally or figuratively, impair others' ability to breathe and/or to live well? Gaian thinking makes clear that violence done to one individual or species – whether physical, psychological, genetic or economic – ultimately affects all. Hurting oneself is hurting others, hurting others is hurting oneself. So for each of us, living as nonviolently as possible makes a positive contribution to the life of the whole. The personal imperative to do so faces us in the suffering of those humans most clearly affected by violence – the starving, the wounded, the exploited, the homeless and the dispossessed.[8]

Therefore: What we look for is not a shift of care from humans to nonhumans, but an extension and deepening of care.[9]

This broadening and deepening the scope of our concern is a positive acknowledgment of the fact that while we are essentially united with our own species by virtue of being human (*what we are*), we are united with all living beings within Gaia by virtue of our existence (*that we are*). Up to now, *what* we are has taken and still takes priority over our existential unity with other species. Now, however, the *way* in which we live out our emerging understanding of that mutualistic unity between our own and other species is becoming more and more important. For Gaia's life-giving, regenerative capacity ultimately belongs to no one individual, group or species but to the sacred community of Gaia as a whole.

Its very sacredness resides in its capacity to give life to each individual being within that whole. Human needs and those of all on the planet are held in balance within Gaia's constant self-renewing. Our attitude to and monitoring of our activities now appear crucial in holding that sacred balance. And we must attend to it if for no other reason than the knowledge that damaging the resources of this community diminishes our own existence.[10]

Present-day mechanical and computer modellings of reality are necessarily statistical and so omit the fact that the character of the whole infuses each single part. And equally, that as whole and part are *internally* related, the whole could not exist apart from the fragment.[11] So as we are shaped bodily, mentally and spiritually by the interdependent character of our participation in Gaia, the character of each of our lives contributes to its character, for good or ill. And so contributes to each others' lives. Our inescapable reliance on those shared contributions makes us vulnerable to changes in them – whether physical, cultural, environmental, social or psychological – and in their ability to sustain us or maintain the quality of our lives.

This truth about ourselves has become ever clearer in the light of climate change events and predictions. Therefore committing ourselves to a nonviolent lifestyle, with all that that entails, is a proper response to a mutual interdependence characterized by what Catherine Keller calls "our Mutually Assured Vulnerability".[12] This makes all living beings vulnerable to the adverse effects on Gaia of increased carbon emissions, industrial waste, mechanized farming practices, depleted fresh water resources and temperature rises. But those persons and species that are powerless to avoid or ameliorate these effects are the most vulnerable of all.

Classic responses to our mutually assured vulnerability

There have been two very different, almost contemporaneous historic responses to this vulnerability. Catherine Keller, using capital letters to highlight its "Mutually Assured" character, takes us back to the era of the Cold War. Then military doctrines of both East and West were based on the necessity of nuclear weapons parity precisely *because* of that mutual vulnerability. It meant those weapons would *ensure* Mutual Destruction of their strategic forces. The clear moral deficit of this MAD (Mutually Assured Destruction) doctrine was that persons and places other than the strategic forces – known as collateral damage or soft targets – would also be destroyed. This led to its being reclassified into "Star Wars" and now into what is called a "Missile Defense Program". These indiscriminately destructive missiles are presented as "protecting" vulnerable NATO countries from attack by "weapons of mass destruction" that might come from North Korea, Iraq or Iran. In every case, this MAD logic requires us to accept that "awareness of the vulnerability of some lives" provides an argument for destroying others. It also assumes that the lives destroyed by or on one side are unrelated to those on the other side: other than as a threat.

Such a doctrine (for so it is preached and held) must be judged not only for its readiness to destroy all living beings within a certain area – on the basis of its supposed isolation from all other areas. Growing awareness of the interconnectedness of global climate change exposes this fallacy. It must now also be judged on the basis of the human and ecological resources committed to the continuing growth of the arms trade and to its industrial and operational impact on communities and environments worldwide. To which we may add the impact of cultural and religious legitimization of military violence targeted at particular ethnic and religious groups and their homelands.

Nonviolent responses to our mutual vulnerability

The second historic response to an awareness of our mutually assured vulnerability was and is also a response to the destructive power of violent racism and military violence. It is the doctrine of nonviolence articulated and lived out by Gandhi to counter apartheid in South Africa and the military violence of the British Empire in India. Arne Næss bases his concept and practice of nonviolent communication and ecological wisdom on this Gandhian doctrine. It holds that violence in any form and coercion of one's fellow humans are indications of moral poverty. This insight is common to the teaching of prophets, philosophers and wise men and women as far back as historical records go. The Buddha and Jesus taught the same lesson. For them as for us, the practice of nonviolence still begins with each individual. But now the idea of Gaia reinforces the message that violence done to one, or by one, ultimately affects all. As does the practice of nonviolence.

Therefore we need to apply the Gandhian term for nonviolence (*ahimsa*) as widely as possible. While philologically it means absence of harm (*himsa)* "to anything on earth in thought, word or deed", Gandhi's use of it goes beyond the predominantly physical in two particulars that are important in our present situation. First, he uses it to designate his ethics of group struggle. Secondly, he applies it to actions or practice in accordance with the root meaning of *ahimsa*, adopting a wide interpretation of "injury" as meaning "*not* avoiding injury to *at least one thing* on earth, in thought, word or deed".[13]

While many of us wish to adopt this wide interpretation and the lifestyle it requires, we also realize its cost. For to travel the path of nonviolence makes us even more vulnerable to the violent forces that dominate nearly all of our societies. This has been and is borne out publicly by the experience of many nonviolent groups and individuals: Gandhi and his followers protesting against the British Salt Tax; Martin Luther King protesting against racial discrimination; to the Greenham Women protesting against nuclear weapons; to those in anti-globalisation/industrialization rallies and to Buddhist monks in Tibet and Burma.

It is, however, encouraging that six Greenpeace protesters arrested for demonstrating against a proposed new coal-fired station in England were acquitted by a jury who accepted their defence of averting the greater violence that would be done by increasing carbon emissions. Wherever recorded, such a response exemplifies Gandhi's dictum that nonviolence is not merely a personal virtue but a social one. It always affects not only the individual embodying it but also those others living in that environment and indeed, the environment itself.

As science now alerts us to and informs us about this interdependence, we find it was already taken for granted in religious expositions of the unity that has always existed between God and the whole community of life. In Genesis 4:7 the Hebrew word *hatta'th,* commonly translated as "sin" is used – *for the first time* – when God warns Cain that it (sin) is crouching at the door like a wild animal, waiting to attack him. He succumbs to the attack, for then "Cain rose up against his brother Abel and killed him" (Genesis 4:8).

Up to now, that stark account has been taken as the central scene in this biblical tragedy. But some scholars believe that by cutting the descriptive elements of the murder down to a bare minimum, the central focus was shifted away from the following scene where Yahweh confronts the murderer with this question:

What have you done? The voice of your brother's blood cries to me from the ground. And now you are cursed from the ground which has opened its mouth to receive your brother's blood from your hand. When you serve the ground, it shall no longer yield its strength to you. (Genesis 4:10-12)

At first it might appear that there are only two actors, Yahweh and Cain, in this drama. But the pointed reference to "the ground" (*adamah*), the fertile soil which alone can sustain life, shows that Earth itself is a third essential actor.

This *adamah* is the realm where human life and work take place (Genesis 2:5). At the same time it is the stuff from which all animals and humans are made and to which they will ultimately return. It is the stage for all the great cycles of life, death, decomposition, decay and new life. The *adamah* appears here as a living organism which has "opened her mouth" and gulped the blood of Abel down her throat (the identical phrase is used in Numbers 16:30, 32 and Deuteronomy 11:6).[14]

Wittenberg remarks that to the Western mindset, that considers the Earth mainly as the material for ruthless exploitation irrespective of consequences, the idea of Earth as a living organism that has the capacity to deny arable soil its power to produce belongs to "the realm of magic". Yet, with the acceptance of Lovelock's idea of Gaia, we are learning to talk about Earth as a living organism with life imbedded in its oceans and soil, covering its surface. Before leaving this drama, it is also worth noting that God expressly forbids anyone to take revenge on Cain. That would simply legitimate extending the impact of his violence onto others.

This takes us back – and brings us forward – to today's context of climate change. When natural disasters such as desertification and loss of fertile soil occur, their effects are exacerbated by already present conflicts that make it more difficult to find solutions acceptable to all. Those rich in resources, for example, have ways of dealing with such disasters not available to the poor; indigenous groups may react violently to refugees. The archetypal tense relationships between Abel the shepherd and Cain the farmer are being played out tragically in Central Africa today. Wherever we live, this divergence in need and response will occur. And it is then that the challenge and difficulties of living nonviolently become apparent.

It is also the time when proven methods of dealing nonviolently with others who have divergent views and agendas become important. They include a willingness to show respect for the views of another person that will always, by definition, remain "other" than our own. Gandhi famously refused to lay charges against a group in

South Africa who beat him up badly for his nonviolent tactics against apartheid. From his hospital bed, he defended their actions as understandable according to *their* culture and ethics; and because he had not explained his own motives and actions clearly enough.[15]

It now becomes clear that the real struggle in increasing awareness of the need for nonviolent responses to the effects of climate change takes place within us: between the urge to own, dominate, control and exploit the other through violent means for our own purposes and profit and the need and desire to live nonviolently; that is, in ways that avoid injury to others as far as possible. And that our moral, imaginative and sympathetic faculties play an important role in living like this. Closely connected to these is the cultivation of a religious imagination that sees reality in terms of our Gaian interdependence and vulnerability; in terms of what is already there.

A nonviolent Christian response

Followers of the Buddha and of Gandhi have kept faith with their nonviolent vision by cultivating their religious imagination. What about the followers of Jesus?

> The "inconvenient truth" of the troubling realities of climate change ought to alert theologians to the "inconvenient truth" that certain readings of sacred texts, and traditional images based on them, have both provided and sanctioned images of God that have in turn sanctioned the violence of Christians.[16]

A nonviolent Christian response to the troubling realities of climate change calls for a move away from images of God that sanction violence. Seen in this light, the life and teachings of Dietrich Bonhoeffer offer us a theological basis on which to build such a response. The original vision of Jesus given him by his clerical, German theological, religious and academic background was challenged by and infused with new meaning through his joining the opposition to church co-operation with Hitler and the National Socialism of the Third Reich.[17] That led to his participation in the development of the Confessing Church and finally, to his imprisonment in 1943 and execution by the Nazis in 1945.

In 1934 he gave a sermon based on Psalm 85:8: "Let me hear what God the Lord will speak, for he will speak peace to his people, to his saints, to those who turn to him in their hearts". In it he dared to hope for what would become a reality in India with Gandhi: winning freedom through nonviolent means. At the same time he exposed the fallacy of Mutually Assured Destruction.

> How does peace come about? Not through a system of treaties or through the movement of international capital … or through universal peaceful rearmament … because here peace is confused with safety … Which of us can say he knows what it might mean for the world if

one nation should meet the aggressor, not with weapons in hand, but praying, defenceless, and, for that very reason protected by "a bulwark never failing"?[18]

During his imprisonment the question addressed at the beginning of this essay: "Who, for me, is God?" became, for Bonhoeffer: "Who really is Jesus Christ for us today?" "Who, or what might be worth living for or indeed, dying for?" After a year in prison his answer became:

> For all the promises of God find their "Yes" in him (2 Corinthians 1:20). Surely everything depends on the "in him". We must again and again sink ourselves, at length and in quietness, in the life, sayings, deeds, suffering and death of Jesus in order to know what God promises and fulfils.[19]

Through this "sinking" himself at length and in quietness in the life, sayings, deeds, suffering and death of Jesus he reassessed his vision of the "lordship" of God. He came to see it as the God of the Bible winning "power and space in the world through his weakness". This vision of lordship could only be expressed paradoxically; as that of a "God edged out of the world onto the Cross". Not only did Bonhoeffer view it in this way. He lived out the implications of that vision in ways that continue to compel our respect.

His theology became focussed on two main points. Firstly, he based his hope in God on the example of nonviolent lordship revealed in Jesus, rather than on expectations of a present and a future governed by an invulnerable, imperial divine power. Then and now, as Pallmeyer spells out, this latter image is routinely invoked to legitimate lethal military and economic violence against our fellow human beings and by extension, against other members of the Gaian community of life on earth.[20]

So our distance from Bonhoeffer's situation within a racist totalitarian regime does not absolve us from the question he faced then and we face now:

> How does our distance from Jesus illuminate what he said in a different time and place and under different historical circumstances? And how does Jesus' distance from us illuminate what we must say and do in the importantly different situation in which we find ourselves today?[21]

This brings me to the second focal point in Bonhoeffer's vision of Jesus relevant to our situation today. He anchored it firmly in a positive vision of earth. For him, "the earth deserved to have the man Jesus". In terms of our present knowledge, it would also be true to say that Jesus, like all of us, emerged from within the Gaian community of life on earth and depended, like all of us, on its gifts. However expressed, positively affirming the connection between Jesus and the earth diminishes the distance between him and us. The interrelatedness of our earthly lives also affirms that, as his life and Bonhoeffer's had the power to make a difference to all our lives now, so too do our own lives have the power to affect our common future on earth.

For then and now, the images of God's lordship that dominate religious imagination are those of a divine governor exercising imperial power on earth through a male hierarchy. The image of God this projects is that of an omnipotent, invulnerable being who exercises absolute arbitrary authority over the lives and deaths of all other beings.[22] Such images implicitly and explicitly condone and indeed legitimate violence against other members of the community of life on earth.

Bonhoeffer's theological critique of this patriarchal, hierarchical theology is founded on the paradoxical lordship of God revealed *to and by* Jesus in his birth, life and death under the supreme power of the Roman Empire. That critique was re-informed by Bonhoeffer's life- experiences and culminated, as he said, in a "God edged out of the world onto the cross". This vision of the lordship of God resonates with the primary and boldest expression of the paradoxical nature of Jesus' lordship: Paul's reflection on Christ crucified:

> For God chose what is foolish in the world to shame the wise and what is weak in the world to shame the strong. (1 Corinthians 1:22-27)

Bonhoeffer's theological critique also harked back to the concept of the "lordship" of God expressed in the ultimate conceptual paradox we call "the Incarnation". There childhood, helplessness and lordship are conceptually united "without confusion, without change, without division, without separation". This paradoxical Chalcedonian Christology arose, Bonhoeffer said, at the cradle in Bethlehem.[23] Before Chalcedon, Tertullian defended this paradox as a "necessary indignity" of faith – both for God the object and man the subject. The double shock of the paradox lies in the moral outrage that God, a perfect eternal being (the received opinion about God) should undergo birth and death; the intellectual outrage in that the flesh impaled on the cross is the self-utterance of God. We are touching here on the paradoxical nature of the unity that is already there between God and the whole community of life on earth.

Today Christians have to wrestle with what appears as a further "intellectual outrage" or "necessary indignity" in regard to ourselves. Over the past 150 years, the implications of Darwin's *Origin of Species* have shaken our pride in ourselves as a species with the status of being specially created by God. It is difficult, and for some impossible to accept that if, as is the case, we are part of the animal kingdom, then rather than being "special creations" we are necessarily kin to all organic life forms, whether extant or extinct. We are lineal descendants of those that emerged some four-and-a half billion years ago and, by virtue of the strange property of Earth's habitability, evolved into what we are today.

In particular, with Jesus, Tertullian and Bonhoeffer we are the "offspring of common parents". Positively, this information about our relatedness brings Jesus closer to us and to the mutually assured vulnerability of our lives on earth. It also illuminates further the paradox of his birth, life and death and, by doing so, re-informs us back into the unity that is already there between God and the whole community of life.

Our own experience of vulnerability makes us look for some source of invulnerability extrinsic to our lives: whether in imperial, social, economic or religious centres of power. Instead, Jesus embodies the paradox of a vulnerable divine lordship unrecognizable as such in customary logical or reasonable terms. It is important to note that Paul, the Church Fathers and Bonhoeffer gave expression to the paradox of Jesus's birth, life and death by making statements about it; not by giving explanations for it. For it defies and exceeds all our reason and logic. In Jesus's own words:

> For human beings it is impossible, but with God all things are possible. (Matthew 19:26)

Including the impossible, Caputo quotes Nicholas of Cusa:

> Since nothing is impossible for God, we should look for Him (in whom impossibility is necessity) in those things that are impossible (but happen) in this world.[24]

Such statements flow from the perception and belief that there exists another perception of the world and of God to which the rational scientific mind alone will never find entrance. The deeper significance of the drama of Cain and Abel is a closed book to it. For that drama takes account of a deeper, earthier and more personal sense of connectedness than do laws based on scientific or logical exposition and technological algorithms. The use of paradoxical categories in regard to God and to Jesus of foolishness-wisdom, weakness-power, might-lowliness, earthly-divine calls for a response other than rational or legal satisfaction.

It calls for a re-evaluation, indeed a reversal of accepted, readily understood religious, political, social and economic values. If earth deserved to have Jesus, what does that say about what is considered valuable in the eyes of the world? Or of the Churches? This is what God has chosen to confound and shame the strong. The identification of the earth, the cradle and the Cross with the mystery of God's wisdom and power has decisive consequences for our vision of Jesus and for our knowledge of God. It teaches us that it is not by abstract analyses of the nature of God as "Supreme Being" or "Almighty Lord" that we best come to understand the revelation in and to Jesus of the lordship of God on earth.

It also teaches us to be constantly and critically aware of the function of God-concepts within our own time and place. This is a hallmark of what, since the 1960s, has become known as liberation theologies. These take consistent account of the lives of those "edged out of" the world's concerns: the poor, the weak, the sexually exploited and the politically and racially oppressed. Bonhoeffer's option for "the God edged out of the world onto the cross" anticipates that of those theologians who put the outcast, the poor and the oppressed at the centre of their concerns. As with him, this gives them a particular perspective and commitment to a God of life: to the God of Jesus whom John quotes as saying: "I am come that you may have life and have it more abundantly".[25]

The spirit of earth's liberation theologians echoes from Bonhoeffer's words:

> I am Earth's guest ... Yet I am charged nothing for food and lodging. Therefore I must be thankful for all the good things that happen to me. I must behave as one guest among all those included with me. I must open my heart to the gifts, pain and joys of Earth. For if you do not have the love of your neighbour/enemy/fellow guest in your heart, if you do not inscribe this love into your personal existence, then you are not "being true to the Earth".[26]

This can also be read as a response to Paul's challenge to us: "What have you that you have not received? And if you have received it, why do you not treat it as a gift?" (1 Corinthians 4:7) Doing this means being mindful of how we behave to all Earth's guests and accepting responsibility for our reciprocal relationships with them; opening our hearts to and being thankful for her gifts that we share with our neighbours, designated enemies and all those who depend on her life support systems. Inscribing this gratitude and love into our personal lives means being open and responsive to the indiscriminate generosity of Earth which gives freely and asks for no return. And therefore withholding neither lodging, food or love from those with whom we share her gifts. It also embodies Gandhi's call to *satyagraha*: becoming attached to the truth, to the truth of what we are, by means of love.

For at the heart of a theology informed by awareness of the reality of being alive in a relational world is a deep resonance with Jesus's vision of the unconditional love of God. This is a God who causes the sun to rise on the good and bad and sends rain on the just and unjust (Matthew 5:47-48). It resonates too with Jesus's dictum that if you are offering your gift at the altar and there remember that your brother or sister has something against you, go first and be reconciled with your brother or sister and then come and offer your gift (Matthew 5:21-24).

All this emphasises the truth of our interdependence within Gaia and our shared reliance on her gifts. Our reciprocal relationships with others make us vulnerable to their character and effects – whether physical, environmental, social, economic or psychological. This underpins liberation theologies' critiques of capitalist systems and ideologies of unlimited growth. For they diminish the ability of the whole to sustain and to maintain the quality of individual lives. Our mutual interdependence assures our mutual vulnerability to the effects of others' actions – for good or ill.

Bonhoeffer's vision of God in the world

This highlights another pertinent and even more radical-sounding aspect of Bonhoeffer's vision of the God envisioned in and by Jesus. Building on his concept of "a God edged out of the world onto the cross", he suggests that we must pursue that vision without "a working hypothesis of God". Or rather, without the usual working hypothesis of an all-powerful, omnipotent deity.

> God is powerless and weak in the world, and that is precisely the way, the only way, in which he is with us and helps us. Matthew 8:17 [He took our infirmities and bore our weaknesses, Isaiah 53:2-4] makes it quite clear that Christ does not help us by virtue of his omnipotence but by his weakness and suffering.[27]

For Bonhoeffer, this meant accepting and learning from the fact that he was living in a world that no longer recognized God as an almighty invulnerable power; as a hierarchical, other-worldly force. Instead, we must accept "powerlessness" as the "given" for a true concept of God: one revealed in the birth, life and death of Jesus. And then use that as an epistemological critique of false notions of God.

Then the "God-forsakenness" of Jesus, that hides God from him, becomes Bonhoeffer's paradigm for the "God-forsaken world". "The God who is with us is the God who forsakes us." But this God who has endured forsakenness is the God who is with us: including those who have forsaken God, "even those of no religion".

But it is important to stress that the God who has forsaken Jesus, and us, is the omnipotent, imperial, invulnerable God hymned and prayed to throughout the ages. We too feel "forsaken" when we are left with a "powerless" God, one who does not use or threaten violence on our behalf. Yet it is this very "powerlessness" of God that confers ethical autonomy on us. But it is the autonomy of the powerless, of those the world calls the weak, the foolish. In other words, it is the power of truth, the truth of our mutual vulnerability manifested in nonviolence and love.

For Bonhoeffer, accepting this truth was an irrevocable step into maturity, the "coming of age" of the world. Accepting it meant doing away with "a working hypothesis" of an omnipotent, imperial God who takes away our responsibility for the world and for our conduct in and towards it.

When he used this phrase he was thinking of Kant's formula:

> The Enlightenment is the emergence of humanity from immaturity that we are ourselves responsible for. Immaturity is the incapacity to use one's own reason without the guidance of another person.[28]

The wisdom of the modern world, says Bonhoeffer, does not need what he called "the tutelage of God". When God is dispensed with as a hypothesis, we come to understand very much better the laws under which the world in fact operates. Dispensing with God as a hypothesis is justified intellectually and scientifically by its results. Not only is it justified. At a time of climate change it is essential that a God responsible for efficient causality in matters of science (such as special creation) is done away with as a "working hypothesis". More importantly for Bonhoeffer, it is justified ethically. The true freedom of a Christian accepts the powerlessness of God's action as the condition of possibility for his own.

Again, we are faced with the demand of an ethics that requires us to reinform our thinking about God. It results in a freedom of will that makes us responsible for the world – and now, for tackling what we have made of it. Bonhoeffer wrote from prison:

> Humanity has learned to cope with all questions of importance without recourse to God as a working hypothesis. In questions regarding science, art and even ethics, this has become an understood thing that one scarcely dares to tilt at any more. But for the past hundred years or so it has been increasingly true of religion also: it is becoming evident that everything gets along without "God", and just as well as before. As in the scientific field, so in human affairs generally, what we call "God" is being more and more edged out of life, losing more and more ground.[29]

In these turbulent times of climate change one of the basic theological questions is how this abandonment of the usual working hypothesis of an omnipotent God might and indeed must function. We can be in no doubt as to its necessity and importance. The "inconvenient truth" of the troubling realities of climate change has alerted us to the "inconvenient truth" that traditional images of an invulnerable, almighty God have sanctioned the violence of Christians against each other and against the Earth.

This has a direct bearing on any re-envisioning of the God of Jesus today. Bonhoeffer offers us one that is at odds with the traditional imperial images of an all-powerful male deity associated with redemptive, apocalyptic but always destructive violence powerful enough to defeat our enemies. In one of the last things he wrote he points the way to a different vision:

> Our relation to God is not a "religious" relation to some supreme, almighty, best of all beings – this is not genuine transcendence – but our relation to God is a new life in "being there for others", in participation in the being of Jesus.[30]

We now know, that as he did, we too participate most effectively in that being when we freely, nonviolently contribute to the wellbeing of Gaia.

Notes

1 Lovelock, vii.
2 Smith (1987), 29.
3 Primavesi (2003), 72-75.
4 Thompson, 1.
5 Margulis and Sagan, 28-29.
6 Næss (1989), 28.
7 Næss (1974), 44.

[8] Næss (1974), 44.
[9] Drengson and Devall, 309-311.
[10] Suzuki, 84-85, 240.
[11] Næss (1989), 58-59.
[12] Keller, 131-3.
[13] Næss (1974), 45-46. Original italics.
[14] Wittenberg, 109.
[15] Drengson and Devall, 220.
[16] Nelson-Pallmeyer (2007), 7.
[17] Scholder (1987), 443-458.
[18] Scholder (1987), 237-239.
[19] Smith (1967), 17.
[20] Nelson-Pallmeyer (2005), 139-153.
[21] Caputo (2007), 34.
[22] Primavesi (2000), 121-136.
[23] Robertson, 30-32.
[24] Caputo (2005), 25.
[25] Primavesi (2000), 94-107.
[26] Bonhoeffer (1971), 13.
[27] Smith (1967), 19; Bonhoeffer (2001), 134.
[28] Smith (1967), 77.
[29] Bonhoeffer (2001), 117.
[30] Smith (1967), 20.

References

Bonhoeffer, Dietrich (1971), *Treue zur Welt*, München: Kaiser Verlag.
— (2001), *Letters and Papers from Prison*, London: SCM Press.
Caputo, John Dominic (2005), "The Experience of God and the Axiology of the Impossible", in: K. Hart and B. Taylor, *The Experience of God: A Postmodern Response*, New York: Fordham University Press, 20-42.
— (2007), *What would Jesus Deconstruct? The Good News of Postmodernity for the Church*, Grand Rapids: Baker Academic.
Drengson, Alan and Devall, Bill (eds.) (2008), *Ecology of Wisdom: Writings by Arne Næss*, Berkeley: Counterpoint.
Keller, Catherine (2005), *God and Power: Counter-Apocalyptic Journeys*, Minneapolis: Fortress.
Lovelock, James (1995), *Gaia: A New Look at Life on Earth*, Oxford: OUP.
Margulis, Lynn and Sagan, Dorion (1995), *What is Life?* London: Weidenfeld & Nicholson.
Næss, Arne (1974), *Gandhi and Group Conflict: An Exploration of Satyagraha*, Oslo: Universitetsforlaget.
— (1989), *Ecology, Community and Lifestyle*, Cambridge: Cambridge University Press.
Nelson-Pallmeyer, Jack (2005), *Saving Christianity from Empire*, New York: Continuum.
— (2007), "Another Inconvenient Truth: Violence within the 'Sacred' Texts", *The Fourth R*, Vol. 20, 9-15.
Primavesi, Anne (2000), *Sacred Gaia*, London/New York: Routledge.

— (2003), *Gaia's Gift: Earth, Ourselves and God after Copernicus*, London/New York: Routledge.

Robertson, Edwin H. (ed.) (1973), *True Patriotism*, London: Collins.

Scholder, Klaus (1987), *The Churches and the Third Reich,* Vol. 1, London: SCM Press.

— (1988), *The Churches and the Third Reich,* Vol. 2, London: SCM Press.

Smith, Cyprian (1987), *The Way of Paradox: Spiritual Life as Taught by Meister Eckhart*, London: Darton, Longman and Todd.

Smith, Ronald Gregor (ed.) (1967), *World Come of Age: A Symposium on Dietrich Bonhoeffer*, London: Collins.

Suzuki, David M. (1997), *The Sacred Balance: Rediscovering our Place in Nature*, Vancouver: Greystone Books.

Thompson, John N. (2006), "Ecology: Mutualistic Webs of Species", *Science* 312, 372-373.

Wittenberg, Gunther (2000), "Alienation and 'Emancipation' from the Earth", in: Norman C. Habel and Shirley Wurst (eds.), *The Earth in Genesis*, Vol. 2, Sheffield: Sheffield Academic Press, 105-117.

Anthropogenic Climate Change and the Truthfulness of Trees

Michael Northcott

A man was famous according as he had lifted up axes upon the thick trees. But now they break down the carved work thereof at once with axes and hammers.
(Psalm 74:5-6)

A goodly saying (is) as a goodly tree, its root set firm, its branches reaching into heaven.
(Surah 14:24)

On May 28 2006 a mud volcano known as Lumpur Lapindo, or Lusi, erupted and began throwing out huge quantities of boiling hot mud in East Java. Since then the mud has submerged the homes and businesses of more than 30,000 residents of the nearby town of Sidoarjo. At first the volcano eruption was blamed on an earthquake that took place 250 kilometres from the area two days beforehand. However scientific investigation since the event reveals that the volcano eruption was caused by a blow out at a gas exploration well, Banjar-Panji-1 that was drilled at the site of what is now a volcano by Indonesian oil and gas company Lapindo Brantas. Seismic records reveal that a huge explosion took place in the well the day before the mud eruption started and that this was caused by an influx of fluid and gas into the well. The well bore was insufficiently protected against such an eventuality and pressure in the well caused an underground blow-out which extruded fluid from the well below ground. As this fluid rose it picked up mud and Lumpur Lapindo, the first humanly-made volcanic eruption in planetary history, was born.[1]

Known locally as Lusi, mud from the volcanic eruption is flowing at up to 150,000 cubic meters a day, which is enough to fill 53 Olympic swimming pools.[2] Efforts to stem the flow have failed and the Indonesian government has constructed a series of dams and dykes in an attempt to direct the flow to the sea. But the project is fraught with difficulties. At one point a gas pipeline was ruptured by the flow of hot mud and caused a gigantic gas burn that went on for many weeks. One of the dams also failed and a village some distance from the volcano was inundated in hot mud. The volcanic eruption has an indefinite life span and so the failure of Indonesia's largest oil company to act in a more precautionary way and adequately protect the well bore from a potential blow-out will require the government and local residents and businesses to bear the costs of Lusi for many years to come.

BURNING THE SUBTERRANEAN FOREST

The growing global demand for oil, and the reduced rate of new discoveries, is driving oil companies to look for oil in ever more marginal places and putting at risk not only the lives of other species who inhabit such places but also human beings and ecosystems.[3] Despite the evidence that fossil fuel consumption is setting the planet on course for a global melt down later in the present century there is not a single national leader who has yet suggested that it will be necessary to regulate and reduce the supply and use of fossil fuels at the point of extraction.[4] Instead politicians and economists in countries such as the United States and Malaysia are promising to reduce taxes on gasoline if elected.[5] Meanwhile Canada is beginning to extract oil from the tar sands of Alberta, which, despite the enormous quantities of water and heat required, is "economic" when oil sells above US $50 a barrel. Australian, British and American electricity utility companies are also planning to build new coal-fired power stations, citing the still unproven and very expensive technology of carbon sequestration as the device that could reduce emissions from these plants should this become necessary under future international climate negotiations. These plants are "needed" because the rate of growth of developed country economies in the last ten years has produced rising demand for energy. And as these countries continue to grow the gap between developed and developing, rich and poor, continues also to grow.[6] And under the stewardship of the World Trade Organisation the expanding global movement of traded goods and investment monies has provoked a tripling of the annual growth rate of fossil fuel use from 1% in the 1990s to 3% between 2000 and 2007. Many blame the rising demand for fossil fuels on developing country attempts to catch up with the still growing wealth of developed countries. But most of the recent strong growth in demand for fossil fuels in the last ten years has come from developed countries and not from developing ones.[7]

Fossil fuels are mostly decayed biomass from ancient trees. This fossil fuel store represents the activity of trees and plants and shell fish over millions of years in response to sunlight, turning the excess quantities of carbon in the atmosphere in prehistoric eras into oxygen, while fixing significant quantities of carbon into the earth's crust, so producing a planetary climate beneficent for mammalian life. These creatures in their prehistoric photosynthesis have provided an ancient subsidy to industrial civilisation which has been the occasion of great present wealth but the subsidy has come at a great price. Sieferle insightfully calls the fossil fuel carbon store that industrial humans are heedlessly removing from under the earth and putting into the atmosphere a "subterranean forest".[8] In just one day industrial humans burn approximately 10,000 years of the carboniferous energy fixed into plant matter through photosynthesis. And with the aid of the oil well and the jet plane the melting of the solid carbon of subterranean forest into the thin air of the upper atmosphere can take place in just a few days.[9]

Demand for energy in Indonesia is growing apace, as from other developed and developing nations. And the production of palm oil for biofuel as well as cooking oil is now the single biggest cause of deforestation in Java, Kalimantan and Sumatra.[10] Palm oil has increased three fold in price in the last five years and deforestation in Kalimantan and Sumatra has therefore grown apace. The protected forests of Kalimantan are shrinking rapidly and being degraded by oil palm encroachment according to satellite data.[11] Indonesia's unique rainforests are home to more than 17,500 plants found nowhere else on earth and this unique patrimony is being destroyed forever.[12] And 40,000 traditional medicines are derived and regularly used from Indonesian plants, nearly fifty per cent of them found only in Kalimantan.[13] The principal agents of Indonesian deforestation are large multinational oil palm companies including Cargill, the world's largest economic corporation, ADM-Kuck-Wilmar, the largest manufacturer of biofuels and Synergy Drive which is controlled by the Malaysian government.[14] There are already 22 factories turning palm oil into biodiesel operating in Indonesia and many more are planned. In theory biofuels are supposed to contribute to reduced climate heating from fossil fuels but if emissions from deforestation are included in Indonesia's greenhouse gas inventory Indonesia is the third largest producer of greenhouse gas emissions after the United States and China.[15] Forest burning and clearance puts out 2,563 megatons of CO_2 equivalents every year which is 4 per cent of global emissions from just 0.1 per cent of the surface area of the planet. There are more greenhouse gases in the soils of Sumatra than are put out by all the world's planes, cars and electricity plants in any one year. Just 4 million hectares of Riau Province in Sumatra stores 14.6 gigatons of carbon, or 40 per cent of Indonesia's peatland carbon.[16] Soils are the largest carbon sink on the planet. The quantity of carbon and methane that is stored in the soils exceeds the other large stores in the ocean and above ground biomass. The growing warming of climates in temperate regions turns the soil into a carbon producer instead of a carbon sink because of increased bioactivity provoked by the warmer weather, which in turn provokes more fungal respiration in soils although tropical soils may be rendered less productive because of climate heating, which has troubling implications for forests and food production in the tropics.[17]

Indonesian forests are initially logged for the extraction of the most valuable timber but the majority of the biomass of the forest is burned thus releasing considerable quantities of carbon into the atmosphere as well as polluting the whole region with smoke. And as the soils dry out after the forests are cut they release vast quantities of carbon. The combined greenhouse gas emissions from the burning of virgin forest and drying tropical soils far exceeds the saved emissions from the fossil fuels that biofuel from palm oil will replace.[18]

While forest burning remains far and away the largest source of Indonesia's annual greenhouse gas emissions at 1400 million tons of CO_2 equivalents, with a further 600 million tons released from drying peatlands, the energy sector is also growing in greenhouse gas emissions which stood at 275 million tons in 2007.[19] This figure will triple at present rates of growth to 716 million tons by 2030. On a per person basis Indonesia's greenhouse gas emissions have grown 150 per cent since the 1980s. The large scale and continuing deforestation of Kalimantan and Sumatra, together with fossil fuel emissions, not only contributes to global climate change, and hence to the suffering of innocent farmers and fisherfolk in Asia and Africa. It also gives rise to local forms of environmental exclusion and injustice. These include the displacement of indigenous forest dwellers, the pollution of rivers and streams turned red with the displaced soil where fisherfolk once found a livelihood, and the raising of land temperatures right across the region.

Indonesia is one of the most heavily populated regions of the world and 70 per cent of Indonesians make a living at least in part from forest lands. These lands are governed in large part by customary rights or *adat*. The Indonesian Constitution in principle guarantees the rights of customary law communities but the *orde baru or* New Order inaugurated under the Suharto regime over-rode many of these customary rights.[20] The continued extension of the palm oil estate of Sumatra and Kalimantan has the effect of making homeless many forest dwellers, both human and other than human. Orang utan are endangered in Sumatra because of the extent of the logging. And at the same time the local guardians of the forest while they protest at the destruction are unable to effectively resist it. Thus the patrimony of the rich biofuel and cooking oil producers such as Cargill, Synergy Energy, and the other large plantation companies continues to grow, the poor and wild species are denied justice.

Deforestation also has disproportionate effects on the urban poor since as well as global climate effects, giving rise to flash floods in low lying urban areas, which are often dwelt in by rural-urban migrants and squatters. As the deforested land absorbs less rainfall and the swollen rivers flood cities and towns downstream of the forests. Deforestation also raises air temperatures in towns and cities as the cooling effects of the trees is removed, thus creating a growing demand for air conditioning and hence growing energy use. The temperature in Indonesian cities has risen more than 1 degree Centigrade since 1990. This again has disproportionate effects on the poor who cannot afford air conditioning or ceiling fans.

More troubling still is the effect of raised temperatures from local and global climate change on crop production and ocean fisheries. Rice yields are likely to decline by at least 4 per cent a year, and maize by 40 per cent a year.[21] Inundation of coastal areas will lead to 95 per cent crop reductions in some coastal districts while rising oceans will also inundate present day aquaculture operations. Rising ocean temperatures will have possibly devastating effects on marine biodiversity. Indonesia's coral reefs have already been seriously damaged by the 1997-98 El Nino event that caused coral bleaching to 16 per cent of reefs. The majority of Indonesia's coral reefs are now in poor condition because of coral bleaching and this is seriously affecting the fertility of the ocean in in-shore waters. Indonesian reefs are still home to the highest marine biodiversity on the planet but destructive fishing methods, as well as ocean warming, are both taking their toll.[22] These methods include the use of explosives and poison and, in a cruel twist, a peculiarly fossil fuelled form of fishing which is electric shock. All three of these methods cause mortality to all creatures in the vicinity and undermine not only biodiversity but the viability of fisheries for future fisherfolk.

Finally climate change threatens the growth of air borne and parasitic diseases including malaria and dengue in Indonesia, both of which are occurring at higher altitudes. Rates of infection are also growing, and new more virulent strains of these diseases are also emerging.[23] Mortality from both diseases has been growing in recent years and climate change threatens even more victims of these diseases in the coming decades.

Economic Drivers of Fossilised and Living Forest Destruction

A major cause of anthropogenic climate change is the growth in the global money supply which now exceeds 20 trillion dollars. This money power is placeless because it is de-material. It detaches making from place and participation from consumption so material lives cease to be spiritual. This is why the Hebrew Bible, Christ in the Gospels, and the Qur'an, all condemn usury and money power – money that bears interest is a destructive but living force that competes with the divine ordering of creation, and that disrupts the material, moral and spiritual ordering of human and other than human life. The exponential growth in the money supply in the last forty years has come about principally because the UK, the United States and Sweden, among others, freed the creation of money from the fractional reserve system which linked paper money and bank credits to quantities of gold held by national banks. Interest bearing bank loans are the most significant source of this exponential growth and, with the brakes off private lending, national governments no longer exercise sovereignty over the money supply and hence they are also losing sovereignty over the human and other than human domains. Thus when American and Chinese investors buy shares in the Indonesian oil and gas company that sparked Lusi they are betting on that company finding new oil supplies and encouraging it to continue to expand the frontier of fossil fuel finds, and they are stimulating the ecological destruction of the lands and oceans around these finds since the oil and gas are typically much harder and more costly in every way to extract in the newest fields, as the case of Lusi demonstrates. The Indonesian response to rising demand for energy is further to put

Indonesian common resources in service of the energy economy both by dangerous and risky oil and gas exploration and by turning forests towards biodiesel production and so the limitless money economy continues to expand its conversion of material power and solar power into stored money and wealth.

Lusi is an example of the destructive and often conflictual impacts of the fossil fuel industry on many regions with large oil reserves from Alaska to Iraq and Nigeria. Deforestation is another source of environmental conflict. The response of indigenous forest dwellers to climate change is to defend the trees. Islam has a profound understanding of guardianship as the essence of the relationship between humans and the divine creation. But while this practice of guardianship involves, and arises from, local guardians in place the social processes that destroy forests, and drill for oil and gas, are disruptive of communities of place and their governance of common resources. In Sumatra Muslims have established protests with NGOs such as Greenpeace against deforestation. Greenpeace Indonesia depicts on its website working parties of volunteer women in head-scarves digging and planting on recently deforested lands in order to prevent the soils from drying out and so releasing more carbon. These people are expressing what Muslims and Christians call guardianship of creatures. Guardianship takes customary form in the common property arrangements through which indigenous peoples have managed shared forests, fisheries, rivers and streams for millennia. But the placeless economy of money displaces the common property arrangements of local guardians.

The abstract economy of money does not respect primary heterogenous forests, but prefers monocrop forests it can count. As anthropologist James Scott argues in *Seeing Like a State* the modern forester loves plantations above old growth forests not because they are more efficient in producing lumber.[24] They are not. Divinely evolved forests produce far more lumber than engineered monocrop plantations. They also produce and sustain many other things the accountant fails to count such as beetles, clouds, fish, freshwater invertebrates, forest birds, fruits, nuts, medicinal plants, primates, orchids, oxygen, rainfall, rodents, rivers, streams, and soil stability; and perhaps most importantly of all they help to stabilise regional and global climates by fixing carbon into biomass through photosynthesis. But trees on a plantation are more countable, more amenable to the accounting procedures of modern capitalism, than the wild activities of unmade forests. And that is why plantations are beloved of modern foresters. Walking his serried ranks of trees, and killing all that grows between them, the modern forester can estimate the yearly added value of the forest growth in the form of a money account. The figures in the balance sheet are an idolatrous simulacrum of the generativity of the divinely ordered creation. They put human industriousness in replacing old growth with countable plantations above the divinely ordered but far more productive heterogeneous species world of the ancient forest.

Forests and trees in religious narratives

In contrast to modern forestry practices indigenous peoples typically have names for most of the trees in the forest and find uses, medicinal and otherwise, for around half of them. The first task given to Adam by God is to name creatures according to the Book of Genesis and in this seminal activity indigenous peoples come far closer to the divine calling to Adam and Noah to guard species. Trees, and relationships with trees, have long been considered indicators of the health or otherwise of human society in the world's religious traditions. The Qur'an has hundreds of references to trees, as does the Bible. For the Hebrew Prophets, and for the Prophet Muhammad human relationships with trees are strong indicators of moral character and spiritual orientation. Both Qur'an and Bible narrate the original fall of Adam and Eve from the divine path of righteousness in terms of their relationship to the tree of the knowledge of good and evil in the Garden of Eden. And not only do humans fall from paradise by a tree. They are also said to be redeemed in part through a tree in the Christian tradition, as Christ dies, and rises again, after being nailed to a tree. As Adam refuses to obey God about the tree it is the obedience of Christ in going the way of the tree, which is the tortuous cross, which is the beginning of human redemption according to the Apostle Paul.

The Hebrew Bible compares the just man who rules righteously to a tree that is well planted by the waterside. Trees in the Bible are a profound ecological symbol of the spiritual order of the cosmos which underlies all reality. This is why both traditions see human relationships with trees as indicative of the moral and spiritual character of human society. In the Hebrew Bible cedar trees play a central role in this respect. The Israelites over-used the cedar forests in their quest to build rich temples and palaces and large warships and the prophets suggest that this reflects the pride and hubris of the late Hebrew monarchy. The height of trees comes to symbolise the arrogance and hubris of the people and the prophets suggest that God will one day cut down the proud rulers who have neglected justice and peace and compares them to the cedars of Lebanon and the oaks of Bashan. Isaiah suggests that when the proud empires have fallen then the forests will be glad because the sound of the axe will no longer be heard at the root of the tree. The Hebrew prophets envisage redemption in arboreal terms. Ezekiel, in chapter 37, in particular develops a striking image of the river of God flower from the crystal temple and giving life to all the earth and watering the trees on its banks. The Revelation of St John the Divine at the conclusion of the New Testament speaks of the River of Life flowing down from the heavenly city watering the trees whose leaves are said to be "for the healing of the nations".

The Qur'an indicates that justice is owed not only to human beings but to all created beings since "we did not create the heavens and earth and everything between them, except with truth" (15:85). And the Qur'an suggests that the truth of God is like the branches of trees reaching from the deep structure of the soil of the earth to the heavens above. The Prophet is quoted as saying, "There is none amongst the believers who plants a tree, or sows a seed, and then a bird, or a person, or an animal eats thereof, but is regarded as having given a charitable gift [for which there is great recompense]."[25] As Nomanul Haq observes so highly does the Prophet value cultivation that he suggests that some will beseech God in paradise for permission to cultivate land and their seeds will grow "colossal as mountains".[26] Even were the end of the world to come, the Prophet suggests that the one who has a palm shoot in his hand should still plant it. As Haq observes "this saying accords a prophetic sacrality to all life: the bounty of nature is a good in itself, even at Doom – a good beyond any immediate or conceivable benefits that one may draw from it".[27]

The Qur'anic understanding of the truthfulness of trees finds a central place in Islamic art and architecture. As a visit to the Alhambra in Spain brought home to me recently, the forms of roots, leaves and trees are seminal to Islamic architecture. Many of the interiors of the Alhambra are dominated by great trees in stone, their branches reaching to the heavens, meeting like a tree canopy, and forming wondrous ceiling patterns. The gift of gothic design may be said to be the greatest bequest of Islam and the Ottoman empire to the Christian West. Abbé Suger of Paris in the twelfth century travelled to some of the great mosques of Southern Europe and brought back gothic architectural principles and deployed them in the building of the great cathedrals of Northern France.[28] Similarly art historians have noted the many Islamic influences in the beautiful built environs of Venice.[29]

The wealthy, and previous generations, knew the value of trees and their health-giving benefits. Flying over the environs of Kuala Lumpur, Jakarta or London older established residential quarters have many trees and these are often where the wealthy live and property values remain high. But in low cost housing areas, and newer housing developments, the trees are typically bulldozed and new trees are not planted when the earthworks and buildings are completed. And to see what effect a complete neglect of the health of trees has in this region it is only necessary to visit Luzon in the Philippines which was once a lush tropical island. But after decades of logging and mining the length and breadth of Luzon, today even the biggest rivers run dry at certain times of the year and much of the land is now arid and gives only a poor return to farmers.

The health of the world's trees is also declining because of anthropogenic climate change. 21 million hectares of coniferous trees are now infested with bark beetle because of climate change – coniferous forests store 49% of earth's biomass store of carbon. Similarly tropical forests are suffering from rising land and air temperatures. With a 3 degree temperature rise scientists believe tropical forests will be at risk of drying out and the whole rainforest ecosystem may collapse as the hydrological cycle on which they rely – and by which they draw water from the oceans and turn it into clouds and hence into rivers – will be undercut by the raised air temperature.

THE ERRORS OF CARBON TRADING AND THE TRUTHS OF TREES

Present human solutions to restraining greenhouse gas emissions involve efforts to control emissions via the creation of markets in which emissions and carbon sinks are counted and commodified. The carbon market involves the creation of carbon credits conferred on developing country forest plantations used in the making of biofuels such as charcoal and biodiesel where these are substituted for fossil fuel use. Hence Indonesia in response to the recent UNFCC conference in Bali, convened to begin the quest for a successor treaty to the Kyoto Protocol, responded to the demand of the delegations and world scientists to reduce emissions from its forests by offering to delay harvesting of plantations, not by setting aside the far more significant quantities of carbon in old growth forests. In fact it refused to set aside a single area of old forest from the destructive and predatory demands of its growing oil palm estate.

The way in which Indonesia responds to the commodifiying pressures of the carbon market is indicative of the fraudulent sums that the Clean Development Mechanism and the other carbon markets involve. Recent scholarly investigations of these new markets reveal that most activities that acquire Kyoto credits which are offset against greenhouse gas pollution elsewhere are activities that were being undertaken anyway. Thus 25% of funds in the CDM went to Chinese hydroelectric plants which the Chinese government was building or had planned before the inauguration of the CDM. Similarly investigations of the large market in carbon offsetting reveal that the majority offset projects are highly unreliable as carbon sinks, and are sometimes just paper exercises.[30] The fiction of the carbon market begins with the procedure itself, which requires that the one who acquires the credit has simply to affirm the intention that the activity for which the credit is acquired is undertaken in part to offset carbon. Such an intention can be claimed falsely without any risk of prosecution for fraud. The neoliberal arrangements for trading carbon imperil the planet because they offer no prospect of reliably reducing the industrial burning of living and fossilised trees. The price of carbon has sunk so low in the European Emissions Trading Scheme that it offers weak incentives for increasing fuel efficiency or de-energising, and since its inauguration the energy intensity of European power production has decreased.[31] The falsehoods of markets in carbon rest upon an ideological presumption against political deliberation and regulation of fossil fuel production and forest destruction, and legal restraint of the powerful corporations who are the principal agents of these activities. This critique is reminiscent of Konrad Ott's third suggestion in the present book that "there is a logical superiority of arguing over bargaining".[32] Carbon trading attempts to resolve anthropogenic climate change through rational choice theory rather than discourse ethics. These falsehoods are the precise opposite of the truthfulness of trees, which, as the Qur'an suggests, mirror the logic and pattern of divine truths in their bodily and biological connections between soil and sky.

If trees are a simulacrum of divine truth then deep listening to trees – ecoacoustics – might involve listening to the divine.[33] This is sometimes dismissed by Christians and Muslims as animism. But when indigenous people make instruments from trees and perform the music of the forests with gongs and other Indonesian instruments they are giving voice to the truths of trees. The French philosopher of science Bruno Latour suggests that the ecological crisis arises in part because trees and other species do not have voices in human parliaments.[34] But no one who has walked just before daybreak in a tropical forest could say that it is voiceless. On the contrary the cacophony of the living forest are a wondrous symphony of sound. The problem is rather that we have become deaf to the voices and sounds of the earth even although it is constantly and rhythmically singing the truths that the creator has set into the creation. This is why it was so interesting that when the tragic Tsunami hit the islands of Western Indonesia in 2004 while humans were deaf to the warning signs domestic and wild animals responded to some kind of earth signal and got away before the Tsunami hit. It is most likely electromagnetic signals or "infranoise" acoustical signals to which they respond.[35] And many indigenous people also got away since they remembered the stories of their grandparents that when the sea disappears it is important quickly to move to higher ground. Trees of course cannot walk and nor can concrete hotels on beaches stripped of mangrove swamps on coastlines with degraded or destroyed corals, both of which in the past provided protection from large ocean waves.

Paradoxically many Western consumers are discovering the sensual wonder of rainforest sounds and they are sold to them in the form of world Music and eco-music. It is possible to buy CDs at New Age stores in my home city of Edinburgh that reproduce the sounds of the rainforest. The selling of forest sounds fetishises the forest, preserving the memory of trees to calm the spirits of Western consumers whose own acoustic environment is dominated by cars, electronic entertainment devices and the hum of machinery. This of course is the magic of capitalism which as Marx memorably suggested turns "all that is solid into air". This famous description of capitalism takes on new meaning in the era of climate change as solid forests, above and below grown, are literally turned into polluting atmospheric gases. And as if in an effort to preserve a portion of the solid world for posterity the process of commodification turns the rainforest sounds into a fetish for New Age Western consumers even as their banks and businesses continue to pour money into the tropical timber and palm oil industries and they consume the products in garden furniture and "health" foods.

Scriptural reflection on trees indicates that the ancients knew far better than science-informed moderns about the co-creaturely and somatic interconnections between humans and other creatures. Trees are alive and they live on a time scale that is beyond the ken of individual human generations. I walked some months ago among eucalyptus trees in the "Valley of the Giants" in Tasmania that were seedlings at the time of Christ and substantial trees during the life of the Prophet. These are the greatest and longest lived flowering plants on the planet but the frontier logic of capitalism continues the relentless drive to turn these precious memorials to earlier eras into woodchips and the land on which they majestically stand into commodified space for more tree plantations. And even as they do this Forestry Tasmania and its principal partner, the Gunns timber company, claim they are reducing not increasing Australia's carbon footprint, against the clear scientific data that old forests store more carbon than new plantations.

Trees have no moral standing in the cost benefit sums of the global economy in part because they are perceived as non-reactive and non-sensual entities. In an attempt to establish the legal standing of trees in a United States court, and in particular ancient redwoods in the Californian Sierra Mountains, Christopher Stone argued that they do have moral standing because, though not evidently sensate, they nonetheless express goal-directed behaviour, and provide homes to other species.[36] But perhaps it is the sedentary and stationary character of trees that makes it hard for modern humans, and the mobile economy of money, to perceive them as beings with any claim on human political economy. The time scale of trees puts then at the very edge of human perception. Trees last on a scale that humans do not. Modern humans – trained in the instant just-in-time and "real-time" procedures of modern economics – do not respect such long-enduring beings. It is interesting for example that when J. R. R. Tolkien turns trees into mythic beings capable of resisting the depredations of Mordor, and the turning of trees into fuel for furnaces fashioning robotic Orks and weapons of war, he makes them mobile. Ents walk and only walking trees can take on the evil empire. Perhaps this is why, when Stone proposed that trees should have legal standing, this was not granted by the courts in California.[37] Standing is not enough. Movement, and preferably at speed, is central to modern canons of respect and worth.

Ecologist David Abram suggests we need to recover an empathic relationship with trees.[38] Again here the scriptures seem to anticipate the attempts at recovery that environmentalists and ecologists foster. In the Hebrew Bible trees clap their hands, they give praise to the creator, they celebrate when the empire falls and the sound of the axe is no longer heard in their midst.[39] Islam too celebrates the co-creatureliness of humans and other species and in Indonesia an environmentalism that arises from Islamic sources is likely to have more cultural power than Western or Christian influenced environmental activism. While in much of the Muslim world, and especially the oil-owning Muslim world, attitudes to the environment remain instrumental at best, and highly destructive at worst, there is nonetheless an emergent environmental resistance among a minority of Muslims to the current dominance of the technologically driven industrial civilisation. A field study of Muslims in Cairo reveals the emergence of a range of pro-environment behaviours such as driving less and recycling in the home, and teaching children the value of cleanliness, which articulate in daily life the messages of the Prophet about the avoidance of waste and gratitude for the fecundity of the divine gift of creation.[40]

The best known advocate of respect for the environment in contemporary Islam is the Sufi-influenced Iranian theologian Seyyed Hossein Nasr. For Nasr God is immanent throughout the material environment and the visible world is a physical manifestation of a much larger spiritual and transcendent order of reality. The neglect of the spiritual and transcendent foundation of all material reality Nasr identifies with the secular materialist cosmology of Western science.[41] For many Muslim intellectuals the answer to this conflict between Western science and technology and the health of the planet is the recovery of a sacred science in which the Qur'an is seen as the fountain of both scientific and sacred knowledge.[42]

NASA's leading climate scientist James Hansen gave a speech to Congress in June 2008, twenty years after another speech to Congress in which he first warned of the coming cataclysm of global warming. He accused the Chief Executive Officers of energy companies of "high crimes against humanity and nature" for funding lies about the science of climate change, and for continuing to extract large quantities of coal, oil and gas without regard for the peril in which they are placing the planet.[43] The clouds and the trees are witnessing to the truth that human beings are changing the climate by their reckless greed and neglect of justice to all God's creatures while oil and gas company CEOs and carbon market makers foster the lie that it is possible to go on with business as usual, expanding the fossil fuelled supply chains of international trade, without harming the present and imperilling the future.

But tropical forests and boreal forests are already showing signs of climate stress, shedding more leaves, bark beetle infested, growth rates declining while those who are driving the destruction of planet earth continue to stop their ears to what the forests already know. Trees cannot help but testify to the truth. They carry in their tree rings an indelible mark of the effects of climate change and indeed the study of fossilised tree rings is one of the most significant sources of historic climate data. Trees cannot lie but humans can and the continuing denial of the connection between deforestation and climate change is indicative of just how.

In the end there will be trees and here the Christian Apocalypse is perhaps uniquely prophetic. If all industrial machinery were turned off, including the electricity supply, the cities of New York and Boston would collapse into the subterranean tunnels on which the tall towers are precariously built. Without air pressure the tunnels would implode within days. And within two or three months tree saplings would begin pushing up between the cracks in the pavements and roads and in the broken and collapsed drains and sewers.[44] Trees last. In a highland glen a hundred miles north of Edinburgh, Scotland, stands a yew tree said to have been alive in the village of Fortingall when Pontius Pilate was borne there. The tree of life in the Garden of Eden is well named. So too is the tree of the knowledge of good and evil. Power-knowledge endangers the morality and spirituality of humans, and now their very survival, because from the human ability to master trees, to cut them down, and to exhume them from the ground, humans fashion empires of power, fuel wars, and create command and control centres so inured to the truths of trees that only an immanent collapse of the earth system can impact on their knowledge gathering procedures.

But so long as trees stand there is hope. And breathing the air of trees, walking among them, waking to the sounds of the species that dwell in them, continue to inspire the guardians of tree and leaf that a different world is possible, and that humans can again become stewards rather than slayers of the divine benison of the forests:

> And he shewed me a pure river of water of life, clear as crystal, proceeding out of the throne of God and of the Lamb. In the midst of the street of it, and on either side of the river, was there the tree of life, which bare twelve manner of fruits, and yielded her fruit every month: and the leaves of the tree were for the healing of the nations. (Revelation 22:21-22)

Notes

1 Davies, Swarbrick, Evans and Huuse.
2 Davies, Swarbrick, Evans and Huuse, 5.
3 Tsoskounoglou, Ayerides, Tritopoulou.
4 Tickell.
5 This promise was made by both Hilary Clinton and John McCain in the primary campaigns of the presidential election of 2008. In Malaysia it was made by Anuar Ibrahim, the country's former finance minister and current opposition leader.
6 UN Department of Social and Economic Affairs, 46-47.
7 Raupach, Marland, Ciais et al.
8 Sieferle.
9 Simms, 95.
10 Sari.
11 Curran, Trigg and McDonald.
12 Koh, Wilcove.
13 Lu.
14 Greenpeace International, 2.
15 Sari.
16 Greenpeace International, 2.
17 Schaphoff, Lucht, Green et. al.

18 Reijnders, Huijbregts.
19 Sari.
20 Colchester, Jiwan, Andiko et. al.
21 Sari.
22 Edinger, Jamaluddin Jompas, Limmon et al., 621-624.
23 Sari.
24 Scott.
25 Haq.
26 Haq, 64.
27 Haq, 65.
28 Bony, 28-29.
29 Howard, 394-395.
30 Haya.
31 Ellerman and Buchner 85.
32 Ott, in this book.
33 Shatin, Topper.
34 Latour.
35 Jagannathan, Makris, 2381; Sheldrake.
36 Stone.
37 Stone.
38 Abrams.
39 Northcott.
40 Rice, 377-379.
41 Seyyed Hossein Nasr, 3-7.
42 Wersal, 453-454.
43 Hansen.
44 Weisman, A. 31-34.

References

Abram, D. (1997), *The Spell of the Sensuous: Perception and Language in the More Than Human World*, New York: Vintage.

Bony, J. (1983), *French Gothic Architecture of the 12th and 13th Centuries*, Berkley: University of California Press.

Colchester, M., Jiwan, N., Andiko, A. et al., (2006), *Promised Land: Palm Oil and Land Acquisition in Indonesia: Implications for Local Communities and Indigenous Peoples*, Moreton-in-Marsh and Bogor West Java: Forest Peoples Programme and Pekumpulan Sawit Watch.

Curran, L.M., Trigg, S.N. and McDonald, A.K. (2004), "Lowland forest loss in protected areas of Borneo", *Science* 303, 13 February, 1000-1002.

Davies, R.J., Swarbrick, R.E., Evans, R.J. and Huuse, M. (2007), "Birth of a Mud Volcano, East Java, 29 May 2006", *GSA Today* 17, February, 4-9.

Edinger, E.N., Jompas, J., Limmon, G.V. et al. (1998), "Reef Degradation and Coral Diversity in Indonesia: Effects of Land-based Pollution, Destructive Fishing Practices and Changes Over Time", *Marine Pollution Bulletin* 36, 617-630.

Ellerman, A.D. and Buchner, B.K. (2007), "The European Union Emissions Trading Scheme: Origins, Allocation, and Early Results", *Review of Environmental Economics and Policy* 1 (Winter), 66-87.

Greenpeace International (2007), *Cooking the Climate,* Amsterdam: Greenpeace International.

Hansen, J. (2008), "Twenty years later: the tipping points near on global warming", *Huffington Post* June 24.

Haq, S.N. (2001), "Islam and ecology: toward retrieval and reconstruction", *Daedalus* 72, 43-76.

Haya, B. (2007), *Failed Mechanism: How the CDM is subsidizing hydro development and harming the Kyoto Protocol,* London: Rivers International.

Howard, D. (2007), "Venice and Islam in the Middle Ages: Some Observations on the Question of Architectural Influence", in: Eva R. Hoffman (ed.), *Late Antique and Medieval Art of the Mediterranean World,* Oxford: Blackwell, 389-404.

Jagannathan, S. and Makris, N. (2005), "Could animals detect the approaching tsunami?" *Journal of the Acoustical Society of America* 117, 2381.

Koh, L.P. and Wilcove, D.S. (2007), "Cashing in palm oil for conservation", *Nature* 448, 30 August, 993-994.

Latour, B. (2004), *Politics of Nature: How to Bring the Sciences Into Democracy,* transl. C. Porter, Cambridge MA: Harvard University Press.

Lu, A. (2008), "Indonesia has 40,000 types of traditional medicine", *China News,* 22 June.

Nasr, S.H. (1997), *Man and Nature: The Spiritual Crisis in Modern Man,* London: Allen and Unwin.

Northcott, M.S. (2007), *A Moral Climate: The Ethics of Global Warming,* London: Darton, Longman and Todd.

Raupach, M., Marland, R.G., Ciais, P. et al. (2007), "Global and regional drivers of accelerating CO_2 emissions", *Proceedings of the National Academy of Sciences* 104, June 12, 10288-10293.

Reijnders, L. and Huijbregts, M.A.J. (2008), "Palm oil and the emission of carbon-based greenhouse gases", *Journal of Cleaner Production* 16, 477-482.

Rice, G. (2006), "Pro-environmental behaviour in Egypt: Is there a role for an Islamic environmental ethic?", *Journal of Business Ethics* 65, 373-390.

Sari, A.P. (ed.) (2007), *Indonesia and Climate Change: Current Status and Policies,* Washington DC:

Pelangi Energi Abadi Citra Environ for the World Bank and DFID.

Shatin, J. and Topper, D. (2004), "Tree Music: Composing with Gaia", *Proceedings of the 2004 Conference on New Interfaces for Musical Expression, Hamamatsu, Japan* at <http: //www.suac.net/NIME/NIME04/paper/NIME04_1D01.pdf>.

Schaphoff, S., Lucht, W., Gerten, D., Sitch, S., Cramer, W. and Prentice, I.C. (2006), Terrestrial biosphere carbon storage under alternative climate projections, *Climatic Change* 74, 97-122.

Scott, J.C. (1998), *Seeing Like a State: How Certain Schemes to Improve the Human Condition Have Failed*, New Haven, CT: Yale University Press.

Sheldrake, R. (2005), "Listen to the animals: Why did so many animals escape December's tsunami?" *The Ecologist* 121 (March).

Sieferle, R. (2001), *The Subterranean Forest: Energy Systems and the Industrial Revolution*, Engl. trans. Michael Osmann, Cambridge: White Horse Press.
Simms, A. (2005), *Ecological Debt: The Health of the Planet and the Wealth of Nations*, London: Pluto Press.
Stone, C. (1974), *Should Trees Have Standing? Towards legal rights for natural objects*, Los Altos, CA: William Kaufman.
Tickell, O. (2008), *Kyoto 2: How to Manage the Global Greenhouse*, London: Zed Books.
Tsoskounoglou, M., Ayerides, G. and Tritopoulou, E. (2008), "The end of cheap oil: Current status and prospects", *Energy Policy* 36, 3797-3806.
United Nations Department of Social and Economic Affairs (2005), *The Inequality Predicament: Report on the World Situation 2005,* New York: United Nations.
Weisman, A. (2007), *The World Without Us,* New York: Picador.
Wersal, L. (1995), "Islam and environmental ethics: tradition responds to contemporary challenges", *Zygon* 30, 451-459.

Climate Research, Interdisciplinarity and the Spirit of Multi-Scalar Thought

Timothy B. Leduc

The October 2008 workshop in Trondheim, Norway, on "Dangerous Environmental and Climate Change" was a rather unique event because it convened a number of scholars to discuss what religious and cultural research can add to academia's global scientific understandings and political economic responses. The need for this expansion of interdisciplinary climate research is epitomized in the uncertainties that have historically afflicted the work of the Intergovernmental Panel on Climate Change (IPCC). As climate physicist Spencer Weart writes, this global interdisciplinary endeavour has significantly increased knowledge of the issue, but has at the same time "scarcely narrowed the range of uncertainty".[1] Explaining the central factor behind this paradox, he goes on to state that predicting climate change requires assessing those rising "CO_2, methane, and other greenhouse gases" that "depend less on geochemistry and biology than on human actions". The implication is that the future climate "actually does depend in part on what we think about it", for our response is deeply intertwined with how we think about human-climate relations.[2] This issue of "what we think" is central to why interdisciplinary research needs a greater inclusion of cultural and religious research, for this recognition makes it clear that the values, beliefs and assumptions underlying scientific and political economic research are deeply intertwined with today's climatic and ecological changes. The two coordinators of the Trondheim event, Dieter Gerten of the Potsdam Institute for Climate Impact Research and Sigurd Bergmann of the European Forum for the Study of Religion and the Environment, in fact brought their respective research orientations together because, as the workshop program states, reductionist science "played a dominant role both in creating regional material wealth and in causing the current dangerous environmental crises". As such, the program also suggested that industrialised "societies are in need of a cultural, or spiritual, revision if the earth is to take a sustainable path into the future". In this chapter, I propose that such a spiritual revision requires transforming the way in which interdisciplinary research thinks about a response to today's climatic and ecological changes. To foster this revisioning process, I will critically assess some historic limitations on interdisciplinary environmental research before contemplating a broader alternative approach that is informed by Western and indigenous perspectives.

In his introduction to the historical evolution of climate science and the IPCC, Weart argues that the research of an "irreducibly complicated" climate system that is interacting with human cultural systems is still too limited by disciplinary specialists who have not been able to adequately intertwine the physical and social sciences.[3] This inadequacy is one reason there has been little effective response to climate change despite the IPCC first concluding in its 1996 Second Assessment Report that the "balance of evidence suggests that there is a discernible human influence on global climate".[4] Offering a way of conceiving these limiting forces is Wolfgang Sachs' elucidation of a "global ecosystems" approach that he argues became dominant after the 1987 Brundtland Commission due to its consistency with international political economic interests. He describes the global ecosystems approach as one which focuses on utilizing ecosystems science in economic plans to sustainably develop a planet "whose stability rests on the equilibrium of its components, like population, technology, resources and environment".[5] As with other environmental analyses[6] and critiques of United Nations environmental pedagogy,[7] Sachs also argued that this approach tended to marginalize the ethical critiques of ecological thinkers like John Muir, Aldo Leopold and E. F. Schumacher. Contrasting these ethical, even spiritual, views that can more strongly confront ideas of sustainable economic development, ecosystems science offered a political economic means to manage nature in a way that would complete "Bacon's vision of dominating nature" while also "manipulating her revenge".[8] With almost 80 percent of the cumulative greenhouse gas emissions since 1990 arising from developed nations like Canada, the United States and the European Union,[9] a case can be made that the IPCC's interdisciplinarity is mired in the Western cultural gravity of such a "global ecosystems" research tendency.

To begin thinking about these trends in academia's interdisciplinary climate research, I want to briefly look at Plato's founding of his school outside Athens' walls in the grove named after the Greek hero Hekademos or Academos. Speculating on the motivation for situating his school in a grove some 2,400 years ago and writing that "a philosopher should 'shield himself behind a wall' ", Robert Pogue Harrison states that perhaps Socrates' condemnation "convinced Plato to seek out a measure of detachment from the polis".[10] This aim for institutional objectivity "was merely a tactical, not ideological, decision", for Hekademos also "had a well-defined political mission: to turn the future counsellors of rulers, if not the future rulers themselves, into philosophers" so that politics "could be grounded upon the foundations of metaphysical truth".[11] Though Plato's original motivation(s) for situating academia in a grove may be impossible to discern with certainty, this strategic positioning did "set a pattern for the future history of academia" as an institution aiming for objectivity on what were once predominantly semi-autonomous garden-like campuses.[12] From the humanistic perspective of Harrison, Western universities have largely remained faithful to Plato's goal of giving "birth to knowledge rather than information, memory rather than mnemonics, and commitment rather than compliance".[13] In contrast, Sachs' definition of a "global ecosystems" trend suggests that when it comes to today's climatic and environmental issues, the interdisciplinary research of Hekademos is positioned too far within the walls of political economic power to faithfully practice such credos. This paper will begin by assessing the validity and implications of Sachs' critique before shifting to humanist, religious and indigenous views on an interdisciplinarity that can be spiritually inspired to re-think, as the Trondheim workshop program put it, "how humans can live in and with the ongoing crisis, and what new orientations could take place".

HEKADEMOS' MULTIDISCIPLINARY WALLS

In an analysis of academia's historic relation to ecological issues, Michael M'Gonigle and Justine Starke write that the ecological crisis is "a crisis of the university" because it is ultimately "both a product of the Western culture that brought on this crisis and an instrument of that culture's self-justification".[14] They begin their look back by highlighting the seventeenth century's enlightened, and ecologically significant, revolution of the West's Scholastic universities, many of which were Platonically-inspired to originate on the edges of great medieval cities like Rome, Florence and Paris. Though deeply informed by Greek "intellectual and philosophical aspirations", the medieval Scholastic tradition's four faculties of law, medicine, theology and liberal arts were initially "enmeshed within the authority and dogma of Christendom".[15] It was only with the sixteenth century Renaissance and following Enlightenment that a major pedagogical paradigm shift occurred as academia became increasingly rationalized. One significant feature in this shift was the view that valid knowledge "required physics and mechanics, not metaphysics and religion".[16] In the place of the Christian focus grew a myriad of rigorously walled disciplines whose greater "conceptual clarity and technical precision" fostered global explorations, merchant trade, economic growth and, what we now know to be, an ecologically and climatically significant Industrial Revolution.[17] With academia today "discovering (but not acknowledging) that the assumed truth of unending growth and progress is also killing the context for all our many truths: the planet," M'Gonigle and Starke propose it is time for universities to undergo another revolution that will allow them to fulfill their role as leading places "where society can think differently, act differently, and can do so right where its citizens live".[18] Long before M'Gonigle and Starke's analysis, David Orr similarly concluded the ecological crisis ultimately "is a crisis *of* education" and that the "discipline-centric education that enabled us to industrialize the Earth will not necessarily help us heal the damage caused by 150 years of industrialization".[19] It is in this culturally challenging context that interdisciplinary environmental research like that of the IPCC has evolved, though, as will be seen, the global ecosystems inertia described by Sachs is a powerful obstacle to this visionary re-thinking of Plato's Hekademos.

Looking at the twentieth century rise of interdisciplinary research across academia, Peter Weingart explains that for a long time it was believed the "unity of science would be achieved at some distant moment in time by way of reducing all fragmented disciplinary knowledge to the fundamental laws of physics", but this belief has given way to a more modest "recombining of different parts of knowledge".[20] The ecological scientist Ann Kinzig proposes "two fundamental reasons" as to why environmental interdisciplinarity has risen in prominence and should continue to be nurtured.[21] First, the potential negative impacts of "environmental problems demands" the active research of comprehensive solutions.[22] Second, if disciplinary knowledge is inherently incomplete because of its reductionism, then the scientific motivation to push "the frontiers of intellectual inquiry compels it".[23] It is in this context of complex ecological issues that seemingly transcend disciplinary research initiatives that the origination of the IPCC can be understood. As Weart clarifies, the IPCC was created in 1988 by the World Meteorological Organization and United Nations environmental agencies to formally coordinate the emerging interdisciplinary research on this complex and potentially dangerous climate issue. Since that time it has coordinated thousands of researchers and governmental delegates from more than a hundred countries, resulting in four successive assessment reports that have increased the certainty that the underlying influence of today's climatic changes are human emissions of greenhouse gases (GHG). The paradox is that this rising certainty about the human role in the climate system is also forcing academia to recognize significant political economic and cultural uncertainties that require the IPCC and interdisciplinary research to be expanded beyond Sachs' global ecosystems approach.

To contemplate the forces that are limiting academia's research of environmental issues, it is helpful to consider the thought of Desmond McNeill and others who differentiate between multidisciplinarity, interdisciplinarity and, as will be discussed later in this chapter, transdisciplinarity. In McNeill's research, multidisciplinarity is described as an approach where the "autonomy of different disciplines" limits the change that can occur "in the existing disciplinary and theoretical structures".[24] Meanwhile, interdisciplinarity offers more flexibility in that it bridges disciplinary terminology and methodologies in cooperative attempts to create "a common framework shared by the disciplines involved".[25] Thinking about these kinds of questions, Julie Klein has argued that effective interdisciplinary research needs to be "supplemented with details of specific connections and more finely structured elements" documented by rigorous disciplinary research, such that "generality and particularity coexist".[26] Stressing this same point, Kinzig proposes that "continued advances in discipline-based science are absolutely essential to maintain and build the foundations upon which interdisciplinary research must be based".[27] In fact, Kinzig proposes that it is time for scientists to become more engaged in interdisciplinary analyses because "scientists can be expected to be more cognizant of the assumptions and limitations of the component pieces" when in dialogue with "policymakers, managers, and other stakeholders".[28] Though this grounding of interdisciplinarity within disciplinary science sounds logical, especially in relation to questions of ecology and climate change, McNeill argues that the institutionally defined disciplinary boundaries of universities can "act as a largely conservative force" for interdisciplinary innovation.[29] This critique that suggests environmental research is to date being predisposed to a narrow multidisciplinary approach can be seen supported in the history of conservation research.

In an analysis of four succeeding approaches to conserving the nineteenth and twentieth century California fishery, Arthur McEvoy concludes that the repeated failure of conservation initiatives were related to the plans being too narrowly focused on prevailing academic paradigms and political economic interests.[30] He describes the twentieth century as beginning with a "laissez-faire" approach of naturalizing the market's exploitation of ecology; followed by the mid-century Progressive Era that focused on scientifically determining the limits of a sustainable yield; moving to the "tragedy of the commons" period when nature was to be internalized into the economy; and, most recently, arriving at an interactive view of human-ecology relations within an economic frame of reference, such as with sustainable development. Offering a complementary historic view on the failure of Canada's east coast fisheries, conservationist Raymond Rogers holds that the attempt of conservation science "to represent natural communities in the profoundly impoverished terms of production and exploitation" reflected political economic assumptions that eventually insured the destruction of both ecological and human communities "within the hegemony of modern economy".[31] Summarizing Rogers' research, ecofeminist Val Plumwood highlights a few telling factors that interacted in the collapse: "the political objective was the maximization of present values;" conservation science was unable "to speak unambiguously for limiting the catch; marine science continually overestimated the resilience of stocks; and from different kinds of scientific advice that were available, those voices that suited the desire to continue fishing at high levels were selected and heard".[32] With the focus on global economic growth manifesting more and more ecological crises, Rogers writes that conservation research continues to be narrowed by being placed "in a constant state of crisis management in which biophysical analysis and policy decisions are cobbled together in an atmosphere of ecological and economic crisis".[33] Following the thought of Rogers, Plumwood concludes that the lack of conservation success over the twentieth century is based in a "rationalist failure to situate the human in ecologically embodied and socially embedded ways".[34] This same historic recognition made it apparent to McEvoy that the research "of environmental change should account for the mutually constitutive nature of ecology, production, and cognition," for excluding "any of the three elements ... is to miss the crucial fact that human life and thought are embedded in each other and together in the non-human world."[35] Such a mutually constitutive definition of interdisciplinarity goes far beyond the multidisciplinarity of Sachs' global ecosystems approach that still dominates academia's climate and environmental research.

The thought of Orr offers "three primary reasons" as to why such a broadly defined interdisciplinarity is making little "headway in the vast majority of colleges and universities" – "institutions of higher education, full of smart and learned people".[36] First, he makes the case that university curriculum and research is still dominated by fragmentary disciplines whose reductionary focus tends to marginalize analyses of "larger trends and patterns".[37] In relation to dangerous climate change, the largest pattern of uncertainty is, as Weart argues, those cultural behaviours and related assumptions that continue to foster extensive GHG emissions, despite the troubling evidence. These patterns go beyond the research strengths of ecosystems science and political economics, for as Orr points out their ethical, moral and even spiritual dimensions are the substance of humanistic liberal arts studies. The second deeply interrelated limiting influence identified by Orr is the professionalizing of academia toward productive outputs. He argues that the constant focus on productivity not only encourages "conformity with standards set by the elite in a particular field," but also that the "resulting narrowness" has "rendered many unwilling and unfit to ask large and searching questions".[38] In a similar, though more general, critique of academic trends, legal ethicist Deborah Rhode states that the impact "of declining government support and increasing research expenses" has pressured natural science researchers "to focus on what corporate sponsors will subsidize", while universities have been lured by the potential revenues of "patents or partial ownership interests".[39] As with Orr's concern about corruptive political economic influences, Rhode argues that though "the benefits of profit-generating ventures are immediate and quantifiable," the more "diffuse and cumulative" risk "is that this short-term commercial agenda will distort scholarly priorities," thus calling into question the "academy's reputation for disinterested inquiry".[40] Supporting this economizing and disciplining of academia are, in Orr's analysis, the universities' administrative leaders who have largely stopped questioning "foundational assumptions of higher education", including "the disciplinary structure of knowledge, the growing dependence of higher education on corporate and government funding, or the implicit belief in the human domination of nature".[41] The earlier discussed analysis of Kinzig supports Orr's three concerns, for she argues that it is time for academia to dismantle barriers to interdisciplinary research, "including those of scarce funding, disciplinary institutional traditions and organizational structures, inadequate interdisciplinary training, and insufficient rewards for integrative research".[42]

Though researching a multitude of environmental issues, the above voices seem to commonly suggest academia's economizing and disciplining trends are predisposing environmental and climatic research to settle closer to the multidisciplinary comfort zone of Sachs' global ecosystems approach. After quoting the classic conservation thought of Aldo Leopold on the need to let ecological knowledge change our "intellectual emphasis, loyalties, affections, and convictions", Orr concludes that the ecological crisis highlights the need for academia to break "free of old pedagogical assumptions, of the straitjacket of discipline-centric curriculum".[43] Being one of those thinkers whose ecological, or land ethic has been marginalized as the global ecosystems tendency has ascended, Leopold offers Orr and us some interesting insights on a broadened interdisciplinary environmental research agenda. It is clear from looking at Leopold's published essays and "Sand County Almanac" that his expansion of Western ethics "to include soils, waters, plants, and animals, or collectively: the land"[44] was grounded in the ecological literacy of academic conservation education, ethical discourses of a liberal arts curriculum that was still at his time a core element of higher learning, and an experiential intimacy with natural processes.

The intertwining of these knowledges can be seen displayed in a February entry to Leopold's almanac that begins by speaking of "two spiritual dangers in not owning a farm".[45] His concern is that the urban separation from natural experience can lead people to suppose that "breakfast comes from the grocery" and "heat comes from the furnace".[46] On the farm, Leopold finds such abstract delusions to be confronted in many ways, including a February blizzard that led him to write: "If one has cut, split, hauled, and piled his own good oak, and let his mind work the while, he will remember much about where the heat comes from, and with a wealth of detail denied to those who spend the week end in town astride a radiator".[47] These details arose from the eighty tree rings of the oak that would heat his farm, for the experience inspired a blending of knowledge on the ecological processes and human history that influenced this tree's growth. In his words, as "two piles of sawdust" began accumulating, I sensed the wood was "the integrated transect of a century; that our saw was biting its way, stroke by stroke, decade by decade, into the chronology of a lifetime, written in concentric annual rings".[48] With such an experiential sense of the need to ecologically embed human communities and knowledges it should not be surprising that Leopold also wrote in his land ethic essay that "the key" to successful conservation will be "to quit thinking about land use as solely an economic problem".[49] As will be seen in the next section, ecological thinkers like Leopold have the experiential-based understanding to not only guide environmental research outside the disciplined and economized walls of a global ecosystems approach, but also to help us think about the role climate change processes can play in the interdisciplinary transformation of Hekademos.

MULTI-SCALAR ECOLOGICAL THINKING

An experiential sense of the land and climate as forces that can change and deepen the way we think about human-nature or culture-ecology relations is not foreign to academic conceptions of what constitutes environmental interdisciplinarity. In 1968, a few decades after the writings of Leopold, Canada's first academic environmental program was founded at York University in Toronto. Of relevance to the preceding analysis is an original curriculum document that is concerned with questions of interdisciplinarity, for it offers the interesting description of "environmental thinking" as the "relatively undeveloped" idea in the West that "subject matter has an intrinsic form of thought as well as content, and that the environmental context of substantive concerns may be used to determine form of thought".[50] Considering the relation of this "environmental thinking" to questions of pedagogy and research, the collection of curricula and pedagogical essays goes on to suggest that such an approach may even promote a transdisciplinary understanding that so transforms disciplinary concepts and methods it becomes "difficult if not impossible to identify individually in the transformed approach".[51] Offering a related view, McNeill states that transdisciplinarity is "based on a common theoretical understanding and accompanied by a mutual interpenetration of disciplinary epistemologies".[52] As with Leopold's education by the oak, the pedagogical document argues that there is academic value in shifting our approach to ecology from an object of disciplinary, multidisciplinary and interdisciplinary study to an adjective that can modify "both that which is to be studied and also the manner in which it is to be studied".[53] One of the co-authors of this early pedagogical document who can help us further expand upon the interdisciplinary dimensions of this *ecological* or *climatic thinking* is John Livingston, the eminent Canadian conservationist and co-founder of the Canadian Broadcasting Corporation's "The Nature of Things" who passed away in 2006.

Around the same time that Leopold was being educated by the land to the west of the Great Lakes in Wisconsin, Livingston's life as a conservationist and naturalist was being fostered by childhood ecological experiences in the natural remnants of 1930s Toronto on Lake Ontario's north shore. As a young boy, he realized "that a ravine" unspoiled and near his house "was full of wild animals, especially during spring and fall bird migrations".[54] The ravine path Livingston took towards ecological thinking was not only motivated by his early relations with "hundreds of kinds of birds, plants, butterflies and other forms of life", but also by the shock of having these relations suddenly erased by a sprawling urban development while still young.[55] Later in life he would write that urban centres offer the most stark experience of the separation between human and natural, with the result being socially pervasive "perceptual and conceptual aberrations" that "may be seen as institutionalized delusions".[56] In much the same spirit as Leopold's concern about the missed ecological "details" of city versus farm living, Livingston held that the pervasive technologically-mediated urban life was enfolding society and, consequently, conservation within "its own conceptions of instrumentality, neutrality and purposiveness".[57] As with Sachs' critique, his classic book "The Fallacy of Wildlife Conservation" argues that conservation has become overly rationalized by ideas of stewardship, sustainable harvest and wise-use, all of which primarily define ecology "as a human asset".[58] To transcend these cultural limitations, Livingston concluded it was necessary to recognize that we had not yet "reached the root cause of environmental distress" – the dualistic nature of "Western industrial man".[59] This concern about the West's human-nature dualism led him, as with many others, into a critical history of the role Western religion and the humanities, including the thought of Plato, has played in fostering today's ecologically dangerous and unsustainable way of thinking.

From Hekademos' position between the ancient city reality of Athens and its surrounding Mediterranean ecology, Plato would, in the words of Livingston, bewail "the disappearance of the forests from the slopes of Greek mountains".[60] Despite being "aware of the effects of deforestation", all indications are that his school spent more time philosophizing about Athens' city politics than thinking about the metaphysical and political implications of these ecological changes. Not only was the philosophic focus ecologically limited, Livingston asserts that he influenced the historic succession of anthropocentric Western philosophers who considered humanity to be "the highest product of creation" and nature to be in existence "expressly for his purposes".[61] Many others support Livingston's view on the ecological impacts of a still present, though transmuted, Platonism that canonizes "the subject/object distinction".[62] Writing on the West's historically changing Ideas of Nature, philosopher Raymond Williams concurs with Livingston that the ecological crisis is "not simply the product of modern industry or urbanism", but is deeply rooted in a Greek philosophy that dualistically contrasted "the state of nature and a formed human state with conventions and laws".[63] Holding a similar view, Canadian public intellectual John Ralston Saul interconnects the anthropocentric and elitist tendencies of Plato's duality to both today's "repeated environmental crises" and academia's continued tendency of reducing reality to "a multitude of factors" that neglect "basic questions about the whole".[64] While "we don't expect human relationships to be rational", he states that complex ecological processes are expected to be so because of the dominance of pragmatic views concerned with economic growth, jobs, "feeding the hungry, and other essentials".[65] Conveniently missed in such pragmatism is the possibility that "the short-term, logical, linear, instrumental nature of this central argument may actually be the root cause of the problems".[66] If, as Harrison speculates, Plato's original intent for situating his school outside Athens' walls in a grove was to achieve a political detachment for facilitating the search of metaphysical knowledge, then perhaps his tactical decision to direct that knowledge at political economic leadership had pitfalls that still today constricts academia's interdisciplinary search for adaptive ecological or climatic thinking.

Raising such critical analyses is not meant to suggest any of them are absolutely correct or, more specifically, that Plato's thought only has ecological maladaptive tendencies; rather I am using them to highlight cultural patterns that indicate academia's ecological crisis reaches far back in time, and neither Plato's Hekademos nor, as M'Gonigle and Starke stress, the medieval urban university are detached from today's unsustainable thought. In fact, both Livingston and Williams argue that another significant influence on today's situation is the Scholastic and, more broadly, medieval Christian belief in the dualism of original sin. Both their analyses share much in common with the seminal 1967 article by historian Lynn White, Jr. that argued the Enlightenment's powerful and ecological destructive "marriage between science and technology" was fundamentally grounded in the preceding medieval Christian ethos.[67] Offering a similar line of thought, Livingston argues the Christian placement of fallen humanity between God above and nature below has promoted the conclusion that there is "no conceivable reason for the existence of the blue planet apart from the needs of God and man".[68] Meanwhile, Williams describes the historical displacement of Europe's preceding pagan animation of nature, in the form of spirits, ancestors and gods that could "embody or direct the wind or the sea or the forest or the moon", by a medieval Christian view of nature's diversity as "the revelation of the one true God".[69] Coming forward in time, he asserts that this universalizing approach would provide the cultural groundwork for the Enlightenment's rationalized conception "of a singular Nature" and the uniquely new "abstraction of Man".[70] This objective positioning of humanity over nature provided a fertile cultural ecology in Western academia for science to develop into the experimental analysis and manipulation of natural laws, and economics into the calculation of a self-interested way to consumptive progress.[71] According to Sachs, it is this methodological detachment that has informed the twentieth century conception of "environment" – a cultural view that in the annals of world history has rarely been depicted in terms "more abstract, passive, and void of qualities".[72] Because such critiques suggest deep cultural influences are energizing today's ecological and climatic crisis, it does on the surface seem ironic that these same analyses also suggest academia's crisis in ecological thinking is related to the continued marginalization of humanistic and religious skills for dealing with the metaphysical dimensions of dangerous change.

Looking for some adaptive models from Europe's pre-Enlightenment that could deepen our sense of a contemporary environmental response, Livingston wrote that the conservationist's concern for birds often makes him "recall St. Francis of Assisi who in the early years of the thirteenth century preached to the birds", and whom White held up as a patron saint for Western ecologists.[73] This connecting of modern conservation with medieval nature mysticism was inspired by White's earlier elucidation of an alternative, perhaps more naturally responsive, strain of Christianity that he saw epitomized in St. Francis, the patron saint for Western ecologists. Considering White's view that "Francis tried to depose man from his monarchy over creation and set up a democracy of all God's creatures", Livingston concluded that this Christian mystic "was doing essentially the same thing as the modern environmentalist or naturalist is doing – trying to create a perspective, in nature, for man as a part of nature".[74] It was an experience he knew well, for, as with Leopold, he wrote that there is "a very special, very personal, and at some times highly spiritual quality about the experience of the naturalist".[75] While the mysterious nature of a mystic experience that "cannot be force-fitted into convenient categories" is something Livingston finds "the naturalist knows – profoundly", he also clarifies that because the naturalist uses conventional and scientific knowledge there is in the most basic sense "nothing of the mystical" in the approach he took to nature.[76] Or is there, for in his final book that won Canada's "Governor General Award", Livingston wrote about the need for accessing other "ecological" states of consciousness if the West is to undergo a sustainable cultural change.

Based upon an awareness of ecological science, Livingston's book entitled "Rogue Primate" described various fields of consciousness that in some ways mirror community ecology's nested hierarchical scales. In this ecosystem science, broader ecological scales are generally described as moving over longer timeframes and broader geographies that are the context for smaller scales consisting of shorter timeframes and more localized spaces. To the physics of this ecological hierarchy, Livingston added an animate dimension that suggested the healthy ecological functioning of biological beings, including humans, may require periods of time when "everyday consciousness" is only secondarily focused upon the individual self, and primarily engaged with a "transcendent and participatory" scale of ecological consciousness.[77] He states both human and non-human beings may have the potential to experience other beings "of their own and of neighbouring species not as others but as simultaneous co-existences or co-expressions of that place, perhaps as extensions of themselves".[78] Beyond a species and regional consciousness, Livingston describes the most inclusive form of participation as a biospheric or, in the terms of earth scientist James Lovelock, Gaian "consciousness of entire populations and communities" that builds at critical times and recedes "when recourse to other forms of self is more appropriate".[79]

Offering a resonant view on the potential of global scaled ecological processes to inform our consciousness, the eco-theology of Anne Primavesi that is represented in one of this book's other chapters and in her most recent book "Gaia and Climate Change" suggests that the emergent nature of Gaia's changing climate is like "a global compass whose effects now call for change in us too".[80] Drawing upon the ecosystems science of Lovelock and other Gaian scientists, she writes that Gaia is composed of "different types of entities at shifting scales" whose symbiotic nature "means that each being is generated with, chemically modulated through and sensitive to other modes of being, whatever, whoever and wherever they may be".[81] While the nature mysticism of Livingston in his urbanizing ravine and Leopold on his farm commonly inspire recognition of the need to bring academic knowledge outside this culture's disciplining and economizing walls, Primavesi offers a view on Gaia's climate change as multi-scalar ecological processes that actively call for a transformation in interdisciplinary thought and research.

Such an animated approach to ecology shares much with many indigenous perspectives on culture-ecology relations. The natural historian and interdisciplinary conservationist Paul Shepard follows a parallel path to Livingston and Leopold by arguing that the long Palaeolithic history of hunter-gathering led the human mind to pattern itself after the animals, ecology and climate that were necessary for survival. By integrating ecological science, natural history, anthropology and psychology, he proposed that the surrounding ecology was experienced as "the outward form of the whole space of the mind", with sensory knowledge being our consciousness and the "unseen distance" the unconscious.[82] As with York University's earlier discussed view that the focus of ecological study can methodologically modify interdisciplinary thought, Shepard writes that the ecologically-grounded indigenous approaches of the past and present have the capacity to "carry us beyond ourselves, pursuing the nature of thought as the thought of nature".[83] Unfortunately, many analyses suggest that a full appreciation of such indigenous ways to ecological thinking has, since the Enlightenment, been marginalized from Western research, though this situation has been changing in relation to the research of various environmental issues like climate change.[84]

Of particular relevance to such an analysis is the social critique of Bruno Latour, for he interconnects the historic analyses of Livingston and Williams to this intercultural marginalizing dynamic by highlighting two cultural divisions that were inaugurated during the West's seventeenth century Enlightenment. In much the same spirit as Sachs' delineation of an objective global ecosystems approach, he explains that the first "great historic divide" allowed Western science to separate nature from culture so that its parts could be researched, controlled and managed. From this objective positioning of human over nature, Western science not only developed the discipline to experimentally manipulate natural laws and economically calculate progress, but instituted a second "great divide" that marginalized other cultural knowledges because of their perceived incapacity to objectively separate "what is knowledge from what is Society".[85] What is particularly interesting about this dynamic exclusion of other views is that for Western researchers it has had both external intercultural and internal cultural dimensions, for as the Enlightenment initiated the exclusion of other knowledges it also gradually marginalized its own preceding religious and humanistic ways of thinking about nature. What makes today's dangerous climatic and ecological changes particularly interesting is the way in which they seem to simultaneously call for a broadening of environmental interdisciplinarity and foster a growing appreciation of indigenous knowledges. In Trondheim I talked about the value of one Inuit concept to conceiving interdisciplinary climate research, and will now bring the spirited insights of Livingston and Leopold into dialogue with this northern way of thinking that is to be found beyond a walled-in Hekademos.

SEDNA'S NORTHERN THINKING

After highlighting those social forces that led to the highly objectifying Western concept of "environment," Sachs contrasted it with the continued existence of cultures which view nature as an "active and productive power which permeates every stone or tree, fruit or animal, and sustains them along with the human world".[86] Interestingly, today's turbulent ecological changes have fostered an increasing interaction between these quite distinct worldviews as Western interdisciplinary researchers focus on questions of regional and global adaptation. My research has highlighted this intercultural process in Canada's north as researchers have engaged the traditional ecological knowledge (TEK) of Inuit so as to document changes in a region that some consider the climate system's "canary in the mine".[87] Unfortunately, research indicates this intercultural approach is being limited by dynamics similar to those referred to by Sachs, Livingston, Orr and Latour.[88] For example, years of work with northern indigenous peoples has led anthropologist Julie Cruikshank to argue that TEK is basically a recasting of enlightened Western ideas like "primitive superstition, savage nobility, or ancestral wisdom", all of which fostered research of local indigenous knowledges "as an object for science rather than as a kind of knowledge that could inform science".[89] After documenting this tendency in a story of caribou migration, Cruikshank explains though TEK may be used by researchers concerned with questions of adaptive ecological management the story's content "would be ignored because it confuses rather than confirms familiar" political economic concerns that direct much Western environmental research.[90] In fact, these stories may be marginalized because, as Sachs states, viewing "nature as a living being" often results in attempts "to carefully circumscribe the range of human intervention", especially in an animated context where "a hostile response is to be expected when a critical threshold has been passed".[91] I attempted to circumvent these tendencies by engaging various Inuit in dialogues about today's climatic changes, and then using their concepts and stories as a way of thinking through these changes. By briefly looking at one Inuit concept in relation to the dangerous ecological and climatic changes currently being experienced in the north, I will attempt to clarify the value of spiritually broadening interdisciplinary research that is experientially grounded in the seemingly conscious thought of a multi-scalar ecology.

It was a fitting choice of the workshop organizers to have the artist Hardy Brix present his beautiful photographic "Reflection on Arctic Landscape" before my presentation, for the icy austerity of his pictures that are also displayed in this book introduced the ecological reality that informs the Inuit way of thinking and living in the far north. When one looks at Inuit ethnographies of the twentieth century, the most common conception of nature seems to be the being known as Sedna. In traditional stories still told today, Sedna is described as an Inuit ancestor who became Mother of all the animals after a series of mythic events that led to her sacrifice on the high seas by a father concerned with propitiating a rising storm.[92] Clinging to the boat, the "father cut off her fingers at the first knuckle" and these "became ringed seals", but she hung on and so he cut the second joint and then the third joint with each becoming a different animal of this northern region.[93] It was only with all her fingers loss that she sank to the bottom of the sea and became Sedna, a being who "lives at the bottom of the sea and controls all the animals in her watery domain".[94] As with Sach's assertion that a living nature tends to circumscribe human behaviour, these Inuit stories also describe Sedna as one who is aware of human actions and can take great offence when her rules for northern living are not followed. In the 1920s, an Inuit elder by the name of Ivaluardjuk told the Danish ethnographer Knud Rasmussen that the "greatest peril of life lies in the fact that human food consists entirely of souls".[95] These souls, in his view, "do not perish with the body", and thus must "be propitiated lest they should revenge themselves on us for taking their bodies". As such, people have a responsibility to be aware of their actions, kill carefully so that animals do not suffer and undertake rituals of propitiation. Improper hunting protocols, bragging about a kill, talking while eating or not sharing the catch could, as the elder Ahlooloo reminisced, display a "lack of respect for the animal spirit world" and cause hunger.[96] If this respect for animals is not shown, then Sedna's hair becomes dirty with the spirits of animals who will no longer be released to hunters, with the result being potential starvation. Today's climatic and ecological changes are manifesting just such a dangerous shift in animals for Inuit communities where hunting is still a central part of northern sustenance.

While it would be easy to focus on an overly spiritualized or mythologized sense of Sedna, such a view would be as fraught and narrow as the global ecosystems' marginalization of humanistic and religious discourse. Just as Leopold's essay on a land ethic was grounded in those experiences referred to in The Sand County Almanac, the Inuit myth of Sedna and its social implications are grounded in the ecological intimacy of a hunting way of life. In fact, Sedna is in Inuit thinking classified as an *indweller* that Daniel Merkur describes as numina that manifest "in the miraculous events of the physical world" and are "more closely analogous to Western concepts of the forces of gravity and magnetism than to concepts of angels, demons, and God".[97] Neither supreme beings nor "personifications of natural forces", indwellers are understood to be "powers that constitute nature" as unpredictable beings "that may be benevolent, neutral, or actively malevolent as the mood strikes it".[98] The numinous potential of Sedna can be observed around a northern community like Chesterfield Inlet where I talked with various elders and hunters about the effect of declining water in rivers and lakes on the arctic char population; the seasonal reduction of frozen water on shifting seal species; and the general warmth on insects, birds and mammals, like the grizzly bear, that are migrating north of the southern tree-line.[99]

Not only have other Inuit and northern communities documented similar changes related to a warming climate,[100] but interdisciplinary climate research has also documented these animal changes. The Arctic Climate Impact Assessment, as with the IPCC,[101] reports that northern wildlife will be impacted due to a "loss of habitat and forage bases", "increased risk of disease", "increased pollution", "increased competition as temperate species expand their ranges northward, and impacts due to increased human traffic and development".[102] Research "covering more than 1,700 species globally" indicates species have already shifted their range "6.1 kilometers toward the poles per decade", and spring has arrived "2.3 days earlier per decade".[103] Since such animal shifts are a greater threat to biodiversity "than any other single human impact except land use change",[104] the IPCC has expressed concern that climate change may "enhance or inhibit these losses in biodiversity".[105] As will be seen, the traditional Inuit perspective on the indwelling Sedna suggests this loss of biodiversity is not simply a causal result of unsustainable cultural emissions that are, as Weart concluded, the ultimate source of uncertainty in the evolving climate research.

To think about the value of an indweller concept to environmental interdisciplinarity, it is helpful to review Nurit Bird-David's anthropological research with the hunter-gatherer Nayaka of South India and the resulting analysis of the West's post-Enlightenment rejection of animism. Ever since the theories of thinkers like E. B. Tylor,[106] she explains that Western researchers of indigenous peoples have dedicated much of their time trying to explain why the natural surround was perceived as animate and the historic origin of this "mistaken strategic guess".[107] For Tylor and others, it was only with the Enlightenment that humanity was assumed to have evolved a science and economic self-interest that could rectify this original religious error of interpretation. As with Cruikshank and Latour, Bird-David argues the resulting view of animism and science as "fundamentally antithetical" became a primary reason why indigenous knowledges were seen as invalid by many researchers.[108] Contrasting this traditional Western view on animism, her ethnography led to the view that hunter-gatherers like the Nayaka, or Inuit, "do not first personify other entities and then socialize with them but personify them as, when, and because [they] socialize with them".[109] Instead of knowledge being about "having, acquiring, applying, and improving representations of things in-the-world", animism is concerned with a "relational epistemology" for "developing the skills of being in-the-world with other things, making one's awareness of one's environment and one's self finer, broader, deeper, richer" so one can maintain "relatedness with neighboring others".[110] This animated approach to thinking about the surrounding ecology – what Thomas Heyd refers to in one of this book's chapters as sentient landscape – sounds remarkably similar to the experiences of Leopold on his Wisconsin farm, Livingston in his Toronto ravines and Inuit hunters who are struggling through today's climatic northern changes.

In much the same spirit as Livingston's scales of ecological consciousness and the Inuit view on Sedna's indwelling power, Bird-David also points out that the Nayaka's animistic cosmos is not limited to humans, animals and plants, but also includes cosmological beings known as devaru. She writes that devaru cannot be understood using the term "spirits" or "supernatural beings", but rather you need to think of them as "superpersons" who mediate the relation between the Nayaka and the other beings of their ecology".[111] For example, a hill devaru "objectifies Nayaka relationships with the hill; it makes known the relationship between Nayaka and that hill".[112] This is remarkably similar to the eco-theology of Primavesi, for she writes that being of Irish Catholic descent means her regional sense of Gaia is personified in "the Gaelic figure of the mother-goddess of landscape known as the *Cailleach Bhéara* or Supernatural Female Elder".[113] Just as the Nayaka associate devaru with socially important ecologies, the Cailleach Bhéara is "attached to natural features of the physical landscape" that has a fertile power "as vast and untameable as the wild".[114] As Irish folklorist Gearóid Ó Crualaoich explains, She is a personified "landscape figure" that is associated with the "flocks and herds of beast" throughout Ireland, but is also locally "accessible to human consciousness and human communication" in certain animals and sacred places, such as mountains.[115] By analogy, the ancestral Sedna can be conceived as the indweller of the Hudson Bay's cold waters, weather, coastal lands and animals, a regional relative of Inuit whose predictable patterns and unexpected ways are relationally important to northern living. While these animated cosmologies depict a regional responsiveness that requires particular ethical and spiritual manners for maintaining well-functioning relations, Primavesi highlights the way in which Gaia's climatic changes connects global scale cultural and ecological processes to regional changes like Sedna's animal shifts. During such uncertain times, an Inuit hunter would be pressured to increase consciousness of Sedna's changing ecological movements and their potential human connections. This is exactly what seems to be occurring in the evolution of climate research, and as such there may be some interdisciplinary and transdisciplinary lessons to be learned from following the Inuit hunter's approach to northern ecological thinking.

Prior to conducting research with Inuit, Hugh Brody spent time with the Beaver Indians of northwest Canada and relates a hunting story that is relevant in contemplating an Inuit approach to ecological thinking. While out with a hunting party, Brody became bewildered by the constantly shifting consensus concerning where game might be found. He learned there were subtle variables which were not hard and firm, but rather required hunters to "accept the interconnection of all possible factors, and avoid the mistake of seeking rationally to focus on any one consideration that is held as primary".[116] In the hunt, "there is no step or pause between theory and practice", and consequently every decision is alterable as the hunter "is always ready to change his directions." The planning and control that Brody finds common to Western environmental management "is at odds with this kind of sensitivity" that needs to be aware of the "constant movements of nature, spirits, and human moods." Uncertain ecological situations require thinking about "more information than the conscious mind can hold", for on the hunt "information may not be consistent" such that "wind changes urge a delay" while "news of animal movements suggests all possible speed".[117] Weighing factors for difficult decisions impresses upon one "the irresolvable quality of knowledge", and at these points "some way of processing the facts so that they yield a conclusion" is needed, and this is exactly what ritual practices and dreams offer. It is important to note that these intuitive methodologies are not done in lieu of ecological knowledge, but rather, as Brody writes, are additional tools for "paying the closest and deepest possible attention to the world".[118] This kind of spiritual and knowledgeable participation with the uncertain relational surround seems remarkably similar to Livingston's view that environmental issues highlight the need for consciously participating in the ecology's multi-scalar thought, from Gaia to Sedna and beyond.

Though a deep attentiveness can allow the hunter to successfully track an animal, it is with the following act of killing and eating souls that Sedna's world of greater danger can appear. As Mark Nuttall writes, Inuit hunters recognize they have a responsibility to ensure "animals are killed properly and their meat, bones and hide utilized in ways that will not offend the animal's guardian spirit".[119] To facilitate this offering of respect, Inuit have used elaborate rituals that are performed prior to the hunt and after the animal's death. For example, before cutting up the seal the elder Analok would explain you need "some meltwater from your mouth and pour it into the seal's snout", for it is believed "that a seal refreshed in this manner would be more likely to return again in the form of another seal for another drink".[120] In a review of culturally diverse hunting rituals, anthropologist Roy Rappaport found such experiences of "death and killing" to also be powerful "signs" that hunters ritually use to produce "trance-like states" capable of revealing many co-existent meanings. Such hunting rituals point "to the limitations of signification", and thus "constitute attempts to push past representation in all its forms to naked, immediate existence".[121] He adds that since ecological processes "are in some degree unknown and in even larger degree unpredictable", it may be more adaptive "to drape such processes in supernatural veils than to expose them to the misunderstandings that may be encouraged by empirically accurate but incomplete" science.[122] It is exactly these kinds of misunderstandings that Rappaport argues underlies "species extinctions, atmospheric warming and the social and political disruptions" of a changing world.[123] Something more than a global ecosystems multidisciplinarity focused on ecosystems science and economic models is needed to conceive this co-emerging danger of Sedna's northern and Gaia's global climatic changes. Perhaps the cultural rituals and mythic stories familiar to academia's humanities and religious studies can guide Western research toward transdisciplinary ecological thinking.

In the Inuit past, such climatic animal changes would result in a shaman being called upon to engage Sedna and assess possible human sources of her anger. With a community séance framing the ritual, the shaman would ecstatically travel below the sea to speak with Sedna about why the animals are being kept away from hunters and what kind of confessions are needed. Unfortunately, enacting such northern confessions today would not much impact the present climatic warming that is co-evolving with Sedna's animal changes, for, as Weart pointed out, most of the destructive GHG emissions have arisen from industrialized Western nations. Being situated within the Western context of these environmentally unjust dynamics, Primavesi transforms the Inuit shaman's call for confession to an eco-theological call for her culture to engage "acts of forgiveness".[124] Forgiveness is for her drawn from the New Testament Greek "connotations of 'changing one's mind' ", but in a way that leads to acts which are confessional in their responsiveness to the human and cultural sources of Gaia's climatic changes.[125] While Primavesi argues that this means being committed to "releasing each other from the destructive effects of" participating in GHG emitting "consumerist societies", she also makes the particularly relevant point, especially considering Sachs' global ecosystems critique, that we need to be released "from the mindset of an economic fiscal cycle".[126] Just as with the Inuit relation with a responsive Sedna, she concludes her book by proposing that "if we want Gaia to release us from the effects of what we have done, then we have to retrace our steps and" re-think "the global dominance of the present capitalist economy and its emphasis on transforming nature into short-term monetary profit".[127] Actively responding to these multi-scalar calls for a change in Hekademos' global ecosystems approach will require some kind of confessional act that can allow researchers to regionally experience and transdisciplinary broaden ecological thinking.

Conclusion: A climatic change of Hekademos

Though there is on the surface much in common between the Inuit responsiveness to Sedna and the statement by York University's environmental document that the ecological "context of substantive concerns may be used to determine form of thought", the animated view of the former calls academic ecological thinking toward a much more spirited transdisciplinary direction. The preceding pages have argued that making such a shift will require re-connecting the extensive scientific, economic, environmental policy and natural history knowledges – all today prominent in environmental research programs – with those once core humanistic and religious discourses concerned with metaphysical realities. Such a proposal supports the Trondheim workshop's overarching conclusion concerning a sustainable response to dangerous climate change requiring an analysis of "the IPCC process from religious studies viewpoints." This point was made by Gary Gardner in a 2003 article written for the "State of the World Report", for here he argued that building a grassroots global ecological response will most likely entail a dialogue with the diverse religious worldviews that inform 83 percent of Gaia's human population. Supporting his view was anthropological research that indicated in cases where cultures have managed their ecology sustainably "the credit often goes to religious or ritual representation of resource management".[128] Through highlighting the adaptive value of other cultural stories, concepts and spiritual practices, this paper has likewise shown the anthropological discipline to be vital in the interdisciplinary and transdisciplinary shift of academia's environmental research. In fact, anthropologists like Rappaport, Bird-David and Cruikshank can be seen as significant social scientific guides for the global ecosystems multidisciplinary tendency because of their critical intercultural thought on metaphysical subject matter that academia has largely marginalized from practical concerns in the humanities and religious studies. These disciplines not only help broaden academic thought on the animated dimensions of Gaia, Sedna, Cailleach Bhéara, other multi-scalar consciousnesses or a land ethic, but also offer insight on the adaptive value of blending ecological knowledge with spirited intuitions on these unknowable culture-ecology relations.

It is not only because of their potential adaptive capacity that humanistic and religious understandings need better representation in climate research, for it is also clear that these disciplines may be central to conceiving the continuation of environmentally dangerous cultural beliefs and practices. As many speakers in Trondheim pointed out, it is vital to consider religion in relation to climate change because of both its potential adaptive offerings and its capacity, in the words of the workshop's concluding notes, to "hinder/demotivate creative adaptation", This latter point was clarified in a presentation by Laurel Kearns that mapped the diverse, changing and politically influential Christian responses to climate change in the United States. The negative religious potential that Kearns referred to was epitomized for me two months following Trondheim in a December 8, 2008 picture on the front of the New York Times. Displayed in full colour on page one was the alter of a Pentecostal Christian congregation in Detroit, Michigan that had upon it a choir, a young girl in ritual dance and two large white American-made SUVs on the altar. With the economic crisis hitting the United States hard, the caption stated "praying to save the auto industry". Such a picture brought home for me the point that both in the enlightened West and in many other cultures around the world, spiritual beliefs and practices fuel behaviour that can either facilitate humanity's creative adaptation to today's dangerous changes or continue to nurture climatically naïve and maladaptive ways of living. Focusing on the positive potential of religion, Gardner concluded that common faith teachings on the spiritual health of "living simply" can be used with more recent ecological arguments to "situate consumption in a comprehensive understanding of what it means to be a developed society".[129] Which path Western society takes may be assessed in the near future by looking at our political economic and academic responses for signs that more interdisciplinary and intercultural voices are being included in the climatic thought of this animated multi-scalar reality.

Notes

1 Weart, 191.
2 Weart, 198.
3 Weart, ix.
4 Quoted in Weart, 172.
5 Sachs, 27.
6 For example Plumwood, Rogers (1995), McEvoy.
7 For example Sauvé, Berryman and Brunelle.
8 Sachs, 32.
9 UNFCCC.
10 Harrison, 60, also see Dillon (2003).
11 Harrison, 67.
12 Harrison, 61, also see M'Gonigle and Starke.
13 Harrison, 70.
14 M'Gonigle and Starke, 39.
15 M'Gonigle and Starke, 27.
16 M'Gonigle and Starke, 28.
17 M'Gonigle and Starke, 28.
18 M'Gonigle and Starke, 17.
19 Orr (1995), 43-44.
20 Weingart, 40-41.
21 Kinzig, 710.
22 Kinzig, 710.
23 Kinzig, 710.
24 McNeill, 314.
25 McNeill, 314.
26 Klein, 21.
27 Kinzig, 710.
28 Kinzig, 709.

[29] McNeill, 316.
[30] McEvoy.
[31] Rogers (1995), 121, also see Rogers (1998).
[32] Plumwood, 26.
[33] Rogers (1998), 127-128.
[34] Plumwood, 27.
[35] McEvoy, 300-301.
[36] Orr (1995), 44.
[37] Orr (1995), 44.
[38] Orr (1995), 44.
[39] Rhode, 47.
[40] Rhode, 22, 47.
[41] Orr (1995), 44-45.
[42] Kinzig, 714.
[43] Orr (2004), 33.
[44] Leopold, 239.
[45] Leopold, 6.
[46] Leopold, 6.
[47] Leopold, 7.
[48] Leopold, 10.
[49] Leopold, 262.
[50] Carrothers, Kline and Livingston, Foray 4, 9.
[51] Carrothers, Kline and Livingston, Foray 5, 7.
[52] McNeill, 314.
[53] Carrothers, Kline and Livingston, Preamble 1.
[54] Livingston (2007), 163.
[55] Livingston (2007), 163.
[56] Livingston (1994), 136.
[57] Livingston (1994), 13-134.
[58] Livingston (2007), 166.
[59] Livingston (2007), 166.
[60] Livingston (2007), 298.
[61] Livingston (2007), 298.
[62] Berman, 60, also see Bowler, Primavesi (2000), and Coates.
[63] Williams, 75.
[64] Saul (2001), 45, also see Saul (1995).
[65] Saul (2001), 273.
[66] Saul (2001), 273.
[67] White, 1203.
[68] Livingston (2007), 305.
[69] Williams, 69.
[70] Williams, 75.
[71] Williams, 77.
[72] Sachs, 34.
[73] Livingston (2007), 314.

74 Livingston (2007), 314.
75 Livingston (2007), 73.
76 Livingston (2007), 73.
77 Livingston (1994), 109.
78 Livingston (1994), 113.
79 Livingston (1994), 117.
80 Primavesi (2009), 131.
81 Primavesi (2009), 68.
82 Shepard, 61.
83 Shepard, 63.
84 For example Berkes, Cruikshank, Riedlinger and Berkes.
85 Latour, 99.
86 Sachs, 34.
87 For example Nuttall and Callaghan, Anisimov and Fitzharris.
88 For example Berkes, Cruikshank, Laidler, Leduc (2007), Riedlinger and Berkes.
89 Cruikshank, 21.
90 Cruikshank, 24.
91 Sachs, 34.
92 See Houston, Pelly, Rasmussen.
93 Pelly, 11.
94 Pelly, 12.
95 Bennett and Rowley, 43.
96 Quoted in Bennett and Rowley, 43.
97 Merkur, 255.
98 Pelly, 26.
99 In October 2004, I delivered a two day workshop in Chesterfield Inlet, Nunavut. The workshop was co-facilitated and translated into Inuktitut by Simionie Sammurtok, then President of the local Hunter and Trapper Organization. Participating were A. Tautu, E. Kimmaliardjuk, L. Autut, C. Kriteedhuk, E. Tautu, B. Putulik and M. Amarok – a group of elders, hunters and a Northern Ranger. I discuss my methodological approach and findings in more detail in two articles (Leduc (2007) and (2006)).
100 For example Krupnik and Jolly, Wohlforth.
101 For example Anisimov and Fitzharris.
102 Arctic Climate Impact Assessment, 60.
103 Wohlforth, 244.
104 Zavaleta and Royval, 281.
105 Gitay, Brown, Easterling and Jallow, 25.
106 Tylor.
107 Bird-David, 68.
108 Bird-David, 69.
109 Bird-David, 78.
110 Bird-David, 78.
111 Bird-David, 71, 73.
112 Bird-David, 73.
113 Primavesi (2009), 72.

[114] Primavesi (2009), 72.
[115] Ó Crualaoich, 101-102.
[116] Brody (1981), 37.
[117] Brody (2000), 260.
[118] Brody (2000), 260.
[119] Nuttall, 393.
[120] Cited in Pelly, 60.
[121] Rappaport, 261.
[122] Rappaport, 452.
[123] Rappaport, 452.
[124] Primavesi (2009), 133.
[125] Primavesi (2009), 133.
[126] Primavesi (2009), 133.
[127] Primavesi (2009), 139.
[128] Gardner, 156.
[129] Gardner, 167-168.

References

Anisimov, O. and Fitzharris, B. (2001), "Polar regions (Arctic and Antarctic)", in: *Climate Change 2001: Impacts, Adaptation, and Vulnerability,* contribution of Working Group II to the TAR of the IPCC, Cambridge/New York: Cambridge University Press, 800-841.

Arctic Climate Impact Assessment (2004), *Impacts of a Warming Arctic: Arctic Climate Impact Assessment*, Cambridge/New York: Cambridge University Press.

Bennett, John and Rowley, Susan Diana Mary (2004), *Uqalurait: An Oral History of Nunavut*, Montreal: McGill-Queen's University Press.

Berkes, Fikret (1999), *Sacred Ecology: Traditional Ecological Knowledge and Resource Management*, Philadelphia: Taylor and Francis.

Berman, Morris (1984), *The Reenchantment of the World*, Toronto/New York: Bantam.

Bird-David, Nurit (1999), "Animism Revisited: Personhood, Environment, and Relational Epistemology", *Current Anthropology,* Vol. 40, 67-91.

Bowler, Peter J. (2000), *The Earth Encompassed: A History of the Environmental Sciences*, New York/London: W. W. Norton and Company.

Brody, Hugh (1981), *Maps and Dreams: Indians and the British Columbia Frontier*, Vancouver: Douglas & McIntyre.

— (2000), *The Other Side of Eden: Hunters, Farmers and the Shaping of the World*, Vancouver: Douglas & McIntyre.

Campbell, Joseph (1976), *Primitive Mythology*, Harmondsworth/New York: Penguin Books.

Carrothers, G., Kline, S. and Livingston, J. (1968/1988), *FESKIT: Essays into Environmental Studies: Being Some Interpretations and Amplifications of the FES Curriculum Model*, Toronto: Faculty of Environmental Studies, York University.

Coates, Peter (2005), *Nature: Western Attitudes Since Ancient Times*, Berkeley/Los Angeles/London: University of California Press.

Cruikshank, Julie (2004), "Uses and Abuses of 'Traditional Knowledge': Perspectives from the Yukon Territory", in: D. G. Anderson and M. Nuttall (eds.), *Cultivating Arctic Landscapes:*

Knowing and Managing Animals in the Circumpolar North, New York/Oxford: Berghahn Books, 17-32.

Dillon, John (2003), *The Heirs of Plato: A Study of the Old Academy (347-274 BC)*, Oxford: Clarendon Press.

Gardner, Gary (2003), "Engaging Religion in the Quest for a Sustainable World", in: L. Starke (ed.), *State of the World 2003*, New York: W. W. Norton and Company, 152-175.

Gitay, H., Brown, S., Easterling, W. and Jallow, B. (2001), "Ecosystems and their Goods and Services", in: *Climate Change 2001: Impacts, Adaptation and Vulnerability*, contribution of Working Group II to the TAR of the IPCC, Cambridge/New York: Cambridge University Press, 237-342.

Harrison, Robert P. (2008), *Gardens: An Essay on the Human Condition*, Chicago/London: University of Chicago Press.

Houston, John (2001), *Nuliajuk: Mother of the Sea Beasts*, Halifax: Triad Films.

Kinzig, Ann P. (2001), "Bridging Disciplinary Divides to Address Environmental and Intellectual Challenges", *Ecosystems*, Vol. 4, 709-715.

Klein, Julie T. (2000), "A Conceptual Vocabulary of Interdisciplinary Science", in: P. Weingart and N. Stehr (eds.), *Practising Interdisciplinarity*, Toronto/Buffalo/London: University of Toronto Press, 3-24.

Krupnik, Igor, Dyanna Jolly and Arctic Studies Center (2002), *The Earth is Faster Now: Indigenous Observations of Arctic Environment Change*, Fairbanks: Arctic Research Consortium of the United States.

Laidler, Gita J. (2006), "Inuit and Scientific Perspectives on the Relationship between Sea Ica and Climate Change: The Ideal Complement", *Climatic Change*, Vol. 78, 407-444.

Latour, Bruno (1993), *We have Never been Modern*, Cambridge: Harvard University Press.

Leduc, Timothy B. (2006), "Inuit Economic Adaptations for a Changing Global Climate", *Ecological Economics*, Vol. 60, 1, 27-35.

— (2007), "The Sila Dialogues: Inuit Qaujimajatuqangit and the Spirit of Climate Change", *Climatic Change*, Vol. 85, 237-250.

Leopold, Aldo (1970), *A Sand County Almanac: With Essays on Conservation from Round River*, New York: Ballantine Books.

Livingston, John A. (1994), *Rogue Primate: An Exploration of Human Domestication*, Toronto: Key Porter Books.

— (2007), *The John A. Livingston Reader: The Fallacy of Wildlife Conservation and One Cosmic Instant*, Toronto: McClelland and Stewart.

McEvoy, Arthur (1987), "Toward an Interactive Theory of Nature and Culture: Ecology, Production, and Cognition in the California Fishery Industry", *Environmental Review*, Vol. 11, 4, 289-305.

McNeill, Desmond (1999), "On Interdisciplinary Research: With Particular Reference to the Field of Environment and Development", *Higher Education Quarterly*, Vol. 53, 4, 312-332.

Merkur, Daniel (1991), *Powers Which We Do Not Know: The Gods and Spirits of the Inuit*, Moscow: University of Idaho Press.

M'Gonigle, Michael and Starke, Justine (2006), *Planet U: Sustaining the World, Reinventing the University*, Gabriola Island: New Society Publishers.

Nuttall, Mark (2000), "Indigenous Peoples, Self-determination and the Arctic Environment", in: M. Nuttall and T. V. Callaghan (eds.), *The Arctic: Environment, People, Policy*, Amsterdam: Harwood Academic Publishers, 377-409.

Nuttall, Mark, and Callaghan, Terry V. (2000), "Introduction", in: M. Nuttall and T. V. Callaghan (eds.), *The Arctic: Environment, People, Policy*, Amsterdam: Harwood Academic Publishers, xxv-xxxviii.

Ó Crualaoich, Gearóid (2003), *The Book of The Cailleach: Stories of the Wise-Woman Healer*, Cork: Cork University Press.

Orr, David W. (1995), "Education for the Environment: Higher Education's Challenge of the Next Century", *Change,* May/June, 43-46.

— (2004), *Earth in Mind: On Education, Environment, and the Human Prospect*, Washington/Covelo/London: Island Press.

Pelly, David F. (2001), *Sacred Hunt: A Portrait of the Relationship between Seals and Inuit*, Vancouver: Douglas & McIntyre.

Plumwood, Val (2002), *Environmental Culture: The Ecological Crisis of Reason*, London/New York: Routledge.

Primavesi, Anne (2000), *Sacred Gaia: Holistic Theology and Earth System Science*, London/New York: Routledge.

— (2009), *Gaia and Climate Change: A Theology of Gift Events*, London/New York: Routledge.

Rappaport, Roy A. (1999), *Ritual and Religion in the Making of Humanity*, Cambridge/New York: Cambridge University Press.

Rasmussen, Knud (1969), *Across Arctic America: Narrative of the Fifth Thule Expedition* [Fra Grønland til Stillehavet], New York: Greenwood Press.

Rhode, Deborah L. (2006), *In Pursuit of Knowledge: Scholars, Status, and Academic Culture*, Palo Alto: Stanford University Press.

Riedlinger, D., and Berkes, F. (2001), "Contributions of Traditional Knowledge to Understanding Climate Change in the Canadian Arctic", *Polar Record*, Vol. 37, 203, 315-328.

Rogers, Raymond A. (1995), *The Oceans are Emptying: Fish Wars and Sustainability*, Montreal/New York/London: Black Rose Books.

— (1998), *Solving History: The Challenge of Environmental Activism*, Montreal, New York and London: Black Rose Books.

Sachs, Wolfgang (1992), "Environment", in: W. Sachs (ed.), *The Development Dictionary*, London/New Jersey: Zed Books, 26-37.

Saul, John R. (1995), *The Unconscious Civilization*, Concord: Anansi.

— (2001), *On Equilibrium*, Toronto: Viking Books.

Sauvé, Lucie, Berryman, Tom and Brunelle, Renée (2007), "Three Decades of International Guidelines for Environment-Related Education: A Critical Hermeneutic of the United Nations Discourse", *Canadian Journal of Environmental Education,* Vol. 12, 33-54.

Sauvé, Lucie (1999), "Environmental Education between Modernity and Postmodernity: Searching for an Integrating Educational Framework", *Canadian Journal of Environmental Education,* Vol. 4, 9-35.

Shepard, Paul (1996), *The Only World We've Got: A Paul Shepard Reader*, San Francisco: Sierra Club Books.

Tylor, Edward Burnett (1958), *Primitive Culture*, New York: Harper.

United Nations Framework Convention on Climate Change (2005), *Key GHG Data*, <http://unfccc.int.2860.php>.

Weart, Spencer R. (2003), *The Discovery of Global Warming*, Cambridge: Harvard University Press.

Weingart, Peter (2000), "Interdisciplinarity: The Paradoxical Discourse", in: P. Weingart and N. Stehr (eds.), *Practising Interdisciplinarity*, Toronto, Buffalo and London: University of Toronto Press, 25-41.

White, Jr., Lynn (1967), "The Historical Roots of our Ecologic Crisis", *Science*, Vol. 155, (number 3767), 1203-1207.

Williams, Raymond (1980), *Problems in Materialism and Culture*, London/New York: Verso.

Wohlforth, Charles P. (2004), *The Whale and the Supercomputer: On the Northern Front of Climate Change*, New York: North Point Press.

Zavaleta, Erika S. and Royval, Jennifer L. (2002), "Climate change and the Susceptibility of U.S. Ecosystems to Biological Invasions: Two Cases of Expected Range Expansion", in S. H. Schneider and T. L. Root (eds.), *Wildlife Responses to Climate Change: North American Case Studies*, Washington/Covelo/London: Island Press, 277-342.

THE CULTURAL DIMENSION OF UN/SUSTAINABILITY

DELICATE DISTINCTIONS BETWEEN SOCIETAL SURVIVAL AND COLLAPSE

Davide Brocchi

"Why do some societies make disastrous decisions?" asks the evolutionary biologist Jared Diamond in his most popular book "Collapse".[1] This question is valid for our society too, in spite of its "progress". 48 years after *Silent Spring* by Rachel Carson;[2] 38 years after the UN-Conference on Human Environment in Stockholm; 18 years after the UN-Conference for Development and Environment in Rio de Janeiro, the main indicators of the *global crisis*[3] continue to increase – in some cases exponentially. Seemingly the reality of our society isn't as dynamic as its image: it is time to ask, what hampers and what might promote a change into sustainability. The present chapter provides some inputs for research and a discussion in this direction and classifies them into four categories: complexity, order, dynamics and evolution.

While an *unsustainable development* leads sooner or later into societal failure, the main goal of a *sustainable development* is to avoid dead ends in "social evolution".[4] The central thesis of this text is that, though social and cultural factors play a central role in unsustainable development, they can also be decisive in a development turn towards sustainability.

We could say the same of religion, a fundamental aspect of culture. Although I am not an expert in religious studies, I will make some reflections about the main religion of the culture area in which I grew up, Catholicism. Is it by chance that the causes of the global crisis can be found in societies, which are characterised by their Christian history? Vice versa, can Christianity promote sustainability?

Before a deeper analysis, it is worth to revisiting a concrete case of societal collapse in view of the above mentioned thesis. The tale of the Rapanui people on the Easter Island is particularly important, because it seems to dissolve the complexity of our own global crisis into a lucid, but disquieting metaphor.

Rapanui's lesson

"The overall picture for Easter is the most extreme example of forest destruction in the Pacific, and among the most extreme in the world: the whole forest gone, and all of its tree species extinct", writes Diamond in "Collapse".[5]

Til 1100 CE this island was covered by a subtropical broadleaf forest, with trees of up to 30 meters high. Then the forest and biodiversity decreased ever more. The last medium-sized and large trees were gone by around 1650, 70 years before the landing of the first Western navigator, the Dutch Jacob Roggeveen.

The causes of such a dramatic devastation cannot be natural, as scientific research confirmed. Diamond's thesis is that the inhabitants of the island destroyed the forest by themselves. The Rapanui exhausted the resources of this small island to build more than a thousand of the massive stone statues: the enigmatic *Moai*.

Several clues confirm Diamond's theory. The Rapanui used a lot of wood and ropes to transport the massive statues over distances up to 15 km, between the quarry and the installation-places. The second ecologic impact of the "Moai industry" came through the surplus of food that was necessary to pay the workers. But the expansion of the agriculture at the expense of the forest was caused also by the growing population on the island, that probably exceeded 15,000.[6]

The increasing destruction of the ecosystem and the intensive exploitation of the cultivated area led to a chain reaction. The loss of forest cover caused increased erosion that damaged cultivable land: the Rapanui replaced it through clearing other parts of forest. Without wood it became difficult to overcome the cold winter and to fabricate good canoes for fishing. Due to the lack of food the Rapanui began to eat sea birds.

From the middle of the 17th century the Easter Island's biodiversity rapidly decreased. The ecologic collapse led to a societal collapse: more and more people died of starvation, there were cases of cannibalism, it came to social conflicts and even civil wars. When the navigator Roggeveen landed, he found no more than 3,000 people on the island. They looked "emaciated, anxious and miserable", Captain Cook wrote in 1774.[7] The survivors were decimated in the following century by the diseases introduced by the Whites, or finished by Peruvian enslavement. 1877 the French anthropologist Alphonse Pinart counted only 111 Rapanui individuals on the island.

The Rapanui "had the bad luck among all the peoples of the Pacific area, of living in the most delicate environment and it holds the big risk of an irreversible forest loss".[8] But "bad luck" and "delicate environment" cannot fully explain why a people worked 500 years long on a very small island for destroying its own livelihood so systematically.

What did hinder a development turn into sustainability on the Easter Island? Probably the combined action of three aspects:

a) *Social inequality*. The Rapanui's society was subdivided into elite and common people. The common people were forced to build their houses in the inland, while the headmen and their families were allowed to live nearby the coast, in relatively big houses (*hare paenga*) made of straw, with a basalt floor and a terrace facing the ocean.[9]
The societal development was determined by a minority for the interests of a minority. Why should the elite be interested in a societal change? As the elite was the driving force behind the construction of the Moai, the costs of that development were mainly transferred to the common people. The survival of the common people depended on the headmen, who controlled the whole stockpiling, transport and distribution of food.
b) *Competition*. The island was subdivided into eleven different territories, each one controlled by a headman and his clan. The clans were constantly involved in a competition to achieve the highest social status that was symbolized by the highest Moai.[10] This fight for the social status hindered a perception of "real" problems and a cooperation for their solution. The competition was not only the cause of the "progress" (the statues became bigger and bigger), but also of the growing environmental devastation. So it is not by chance that the biggest Moai was also the last one to be erected.[11]
c) *Ideology*. Without considering the cultural dimension, it becomes very difficult to grasp the "irrationality" behind the construction of more than thousand Moai and the reason why this society sacrificed so many resources and lastly its own existence. The Moai were artistic status symbols that marked the difference between elite and common people. These statues represented the high-ranking ancestors and also had a religious function. This mix of social and religious symbolism mirrored also the link between headmen and priests. They legitimated their power in two ways: through the assumption of a close relationship with the Gods and through several promises to the common people, for example of prosperity and of a rich harvest. This "ideology" was supported by monumental buildings and ceremonies that had to impress the common people.[12] The costs of such "power-demonstrations" were paid by the common people, through the expropriation of the surplus of their food-production.

What happened when the first signs of the crisis became evident? The headmen and the priests convinced the common people, that sacrifices were necessary to appease the ancestors and to get prosperity and good harvests again. Maybe this is the motive for the construction of the enormous Moai after 1600.

The combination of these three characteristics made the social system rigid in the face of necessary changes – and led to its own demise. The reality of a deepening crisis could not be blurred forever: more and more people suffered due to the lack of food. All the sacrifices and prayers to ancestors and Gods brought nothing. What was holy before, became suddenly the target of a deep and diffused aggression.[13] The competition for the best status increasingly turned into armed conflicts and civil wars. Around 1680 a coup by military leaders called *matatoa* overthrew the headmen and the priests and brought a new cult based around a previously unexceptional God *Makemake*. Nonetheless the *tipping point* of the societal collapse had been already passed and was now irreversible. "Not only the old political ideology had been a failure, but also the old religion."[14]

Complexity

The Rapanuis felled all the trees every one, although they needed canoes to leave one of the world's most isolated inhabited islands: "could they really be that stupid?"[15] This question is legitimate, but the "stupidity" isn't only a characteristic of the Rapanuis' people: somewhere everybody suffers from *derealisation*.

Knowledge (of the truth) is a precondition of right behaviour (ethics),[16] but because we cannot know, remember or foresee everything, we make mistakes. Our cognitive capacities are limited: sometimes we do things, we wouldn't do, if we remembered similar mistakes in the past or would know with certainty every future consequence of our behaviour.

The *cognitive limits* belong to the nature of the human being. An important corollary of this assumption is *cultural relativism*:[17] every worldview, religion, theory, value or opinion is only the expression of a *viewpoint* among others. "Relative" doesn't mean "false". A viewpoint is also relative, because it is *related* to a reality but not able to grasp that reality at all.[18] Just as *the map is not the territory*,[19] although it refers to the territory.[20]

How do humans comprehend the complexity in spite of their cognitive limits? By reducing the complexity. A really simple answer, but with dramatic consequences. Our perception of reality is a selective one. Our worldview is a *construction* of reality, a *cognitive map*[21] of the environment.

Humans have an individual will. Johannes Kepler and Galileo Galilei defended their discovery, that the earth rotates around the sun, against the dominant worldview of that time. But nobody can isolate himself completely from the rest of humanity; nobody endures a situation of absolute loneliness. The nature of the human being is a communicative, social and sexual one: relationships among humans are as essential as networks for neurons.[22] Therefore the construction of reality can be only a *social* one.[23]

For the social construction of reality humans need common codes and systems of meanings, that permits us to think about the world and communicate about it with each other. This cognition leads to a semiotic definition of culture as *software of the mind*.[24] Culture is a *collective cognitive map* – but there can be very different maps of the same territory (v. cultural diversity). That every generation inherits a culture from the previous one (everybody has been "born" as Catholic or Muslim and so on...), doesn't exclude a subjective or historic dynamic in the culture.[25]

Every culture makes an environment into a *world*. Paraphrasing Ludwig Wittgenstein,[26] the limits of our world are the limits of our culture. We don't reduce complexity only through interpretation, but also through values and norms that steer our behaviour, the selection of priorities, the development of new technologies, and the use of machines. Culture represents the societal blueprint that runs through societal development.

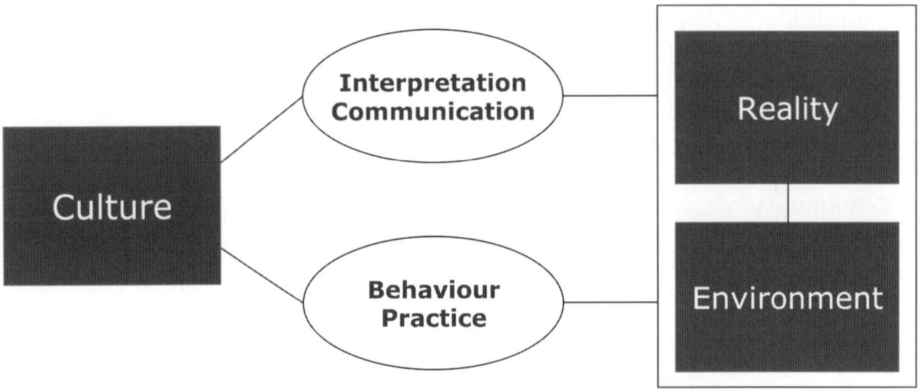

1 The double selective function of culture

Social systems represent a double reduction of complexity on the basis of cultural patterns. For Niklas Luhmann the boundary between social system and environment is a *difference*. This difference depends on three factors: besides the cultural viewpoint (the "observer" in Luhmann's theory) and the cognitive limits there are also *limits of governability*. Humans cannot control complexities. The Rapanui were even overtaxed by self-caused problems. Especially exponential processes and mechanisms of self-alimentation (tipping points) make crises uncontrollable.

The boundary of every social system is the limits of its particularly language, way of thinking, knowledge, resource-availability and management-capacity. We could describe this boundary as a relative point on a continuum, for example between usual and unusual.

The boundary between social system and environment is as relative as the culture, but it marks a distinction between inclusion and exclusion. Outside of our system there is not only an *ecological environment*, but also an *emotional environment* (e.g. the "subconscious", in its deep psychological meaning), a *social environment* (e.g. people we exclude) and a *multicultural environment* (the many cultures that we experience as foreign cultures). A culture behaves towards these environments in the same way.

Social System	Environment
Near	Far
Important	Unimportant
Graspable	Ungraspable
Known	Unknown
Own	Foreign
Trust	Mistrust
Security	Insecurity
Controllable	Uncontrollable
Useful	Useless
Order	Chaos

Table 1: *Boundary between social system and environment*

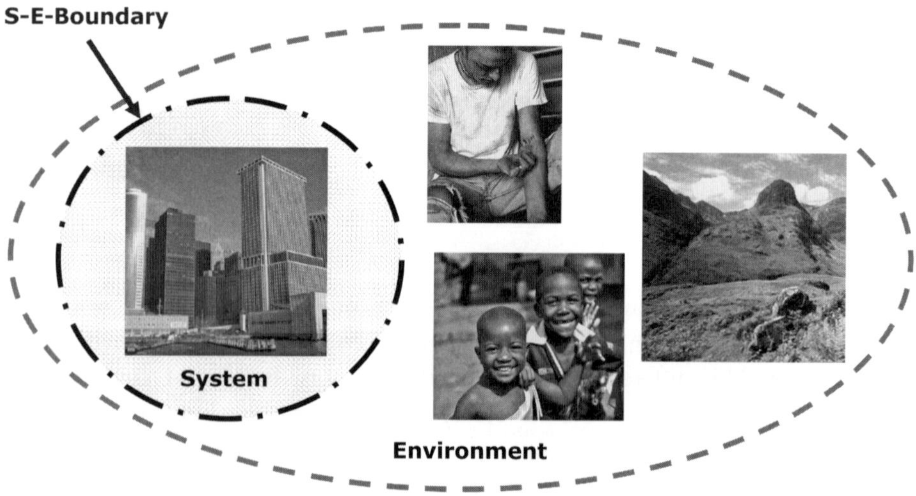

2 *The multidimensionality of the environment*

Which inputs can the perspective above offer to the analysis on un/sustainability? I will resume them in four points.

THE HUMAN LIMITS TO GROWTH

We use everyday words like "globalization" and "nature" as if we could keep such realities in our hand. Governments should guarantee jobs for all, fight terrorism in Afghanistan and tackle at the same time climate change. The truth is: we cannot even comprehend and control the complexity of our own personality, as Sigmund Freud discovered. Therefore we should complete the thesis of Dennis Meadows on *the Biophysical Limits to Growth*[27] with an additional dimension: the *human limits*. Not only the planet is limited, but also the human being, that steers the growth and is steered by it. The human brain is relatively big, nevertheless limited. Individuals cannot "function" in a complex social system and at the same time solve complex environmental problems. Humans aren't Gods, but consist of 70 percent of water. The human being remains limited in spite of cultural differences or the level of societal development – and that could explain why the indigenous society of the Rapanui as well as the modern society isn't immune from ecologic problems.

THE UNIVERSALIZATION OF THE VIEWPOINT

Humans are not able to reach universal truths, but unfortunately they are able to universalize the (own) relative truths. An example: in the western worldview money is indispensable to survive, the modern city is a trusted social system, whereas the tropical forest is a chaotic environment. The perspective of indigenous peoples is probably opposite: while the forest is a trusted system, money is nothing to eat and modern cities stink. Unfortunately the relativity of these two different cultural viewpoints is obscured by the *structures of social inequality*, in particular by hierarchical relationships. Globalization universalizes a relative viewpoint together with the power of (limited) persons and institutions. The habitat of indigenous people is destroyed for the production of goods for the global market.

THE SELF-REFERENTIALITY OF DEVELOPMENT

The material results of development are less relative than the cultural patterns that drive development. The immaterial "walls" in the minds can become physical walls, whose existence is objective and indisputable.

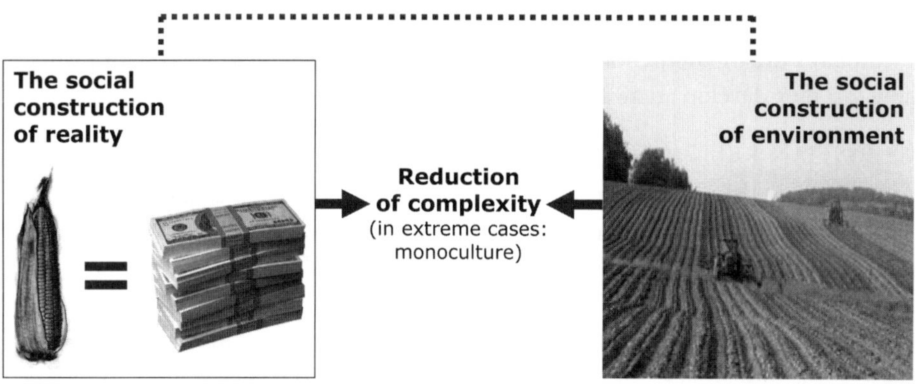

3 The link between culture and societal development

Between the social construction of reality and the social construction of environment there is a continuous interplay: the culture shapes the society that in turn shapes the culture. Pierre Bourdieu would speak of "structured structuring structures",[28] Hartmut Esser of "co-evolution".[29]

In globalization a rational simplification of reality leads to an industrial simplification of environment. While controllable, useful and trusted things are internalised into the social system, disorder, useless things, contradictions and alternatives are banned from the environment.

Globalisation is a form of rationalising development that makes a part of the environment controllable by transforming it into an artificial world that mirrors the dominant worldview.

The assumptions of the founders of modern sciences[30] should be reinterpreted from another viewpoint: only the world that is structured according to mathematics can be explained by mathematics in an exact way.

The self-reflexivity of development

Winston Churchill once said: "We shape our buildings, and afterwards our buildings shape us".[31] We are a part of the same reality and environment which we are interpreting and transforming at the same time. Max Horkheimer wrote that through the rationalisation of society through technologies, not only the external nature of the human being becomes controlled, but also his/her internal nature.[32] The dominance over nature includes intrinsically the dominance over human beings.[33]

The artificialisation of the environment as well as the virtualisation of the reality is changing the human being itself. Generation after generation we are increasingly separating ourselves from nature, making us dependent on artificial things, that were not really missing till some years ago. "If everybody drives a car, I need a car": first the wide spreading (globalization) of "new pseudo-needs" make them into unalienable premises of social integration/identification.

Today's urban children know more heroes from Japanese mangas than plants:[34] how can they develop an awareness of nature?

Christian religion and complexity

According to the sociologist Richard Münch "a distinctive characteristic of the Jewish-Christian religion in comparison to the other world religions isn't only the intellectual rationalization – with roots in Hellenism and in the Enlightenment – but much more the central position of the active designing of the world after religious-cultural ideas (social rationalization)".[35] Is the Jewish-Christian religion the cultural cause of the industrial reduction of complexity and diversity in the modern society?

In the dealing with complexity there are some aspects of the Christian religion that deserve critical attention. The Jewish-Christian tradition extols the "difference" between system and environment with a strong "separation" between Heaven and Earth or body and soul. "Especially in its Western form, Christianity is the most anthropocentric religion the world has seen."[36] The dogma of "papal infallibility" risks making a human into God.

But in Christianity there are also aspects that promote a more open dealing with complexity. In the "Sermon on the Mount"[37] Jesus Christ warned against "false prophets". Many Christians make the emarginated part of the Creation to the centre of their work and mission.

Order

Everyday in the industrialised nations we can find enough food in the supermarkets of our cities; every month we receive wages for our labour; we can trust people that we meet on the street.

The social order is our daily *normality*, but societal crisis show us that such order is nothing obvious. It does not matter if the social order is called globalization or sustainability: the survival of billions of people depends on that, particularly those who live in urban regions and cannot produce food for themselves.

Cultures are operating and normative systems that guarantee that every social order works. They are internalized by every individual through *socialization.*

If culture structures the social order and sets its environmental boundaries; if culture regulates the exchange between social system and environment, then the environmental crisis is a cultural crisis too and needs cultural solutions.[38] If the culture is the blueprint of the society, then the most durable and deep societal change is a cultural one. If every development model is a cultural programme, then *another development*[39] needs *another culture.*

The notion of *culture of sustainability*[40] is a hypothesis of action-research in this direction, as discussed in the following.

Elements of a culture of sustainability

The definition of a culture of sustainability is a challenge – and the motives are different:

- Today what is called "sustainability" is not always sustainable and what contributes to sustainability is not always called so. The same word has different meanings in different sub-cultural contexts.
- A "culture of sustainability" cannot be a list of morals or even of "socially expected answers". More important are the deeper and sometimes unconscious aspects of cultures. But how can such aspects be "rationalized" and "normalized"? Or should a culture of sustainability speak the language of emotions?
- No monoculture is sustainable. Every culture is the expression of a viewpoint – and it is valid also for a "culture of sustainability".

The definition of sustainability shouldn't be reduced to the institutional ones of the *Brundtland Report* (1987) and of the *Agenda 21* (1992).[41] The public and scientific debate on fair and ecological development models is much older. Its institutionalisation through the United Nations isn't necessarily a strengthening, but the result of neutralizing compromises of "United Governments", that do not contain any systemic critic and are often oriented toward an abstract future.[42] The document *What now? Another Development* by the Dag Hammarskjöld Foundation (1975)[43] – for example – gave much more attention to the radical demands from *Eco-Development approaches* and *Dependencia Theories*.[44]

An important novelty of the debate on sustainability is the awareness of the interdependence of topics that were considered separately before (e.g. biodiversity, climate change, work, social justice, democracy, peace, human rights). Therefore a "culture of sustainability" means first to re-define a wide *system* of notions on the basis of their interrelations.

A culture of sustainability is the result of an *enlightenment* from the modern myths of progress, profit, growth, performance, competition, acceleration and private property.

The *change of paradigms* could be oriented to the following arguments:

- *Everything is connected with everything*.[45] The *interconnected thinking*[46] outpaces the Cartesian separation between *res extensa* and *res cogitans*,[47] nature and society, object and subject, external and internal nature of the human being.
  Systemic interactions can be described better through circular processes of feedback than through linear chains of causes and effects. The roots of a culture of sustainability are in the Greek and in the Far Eastern "cultures of circulation", because the ecologic balance is based on circular and not on linear processes.[48]
- *Nature knows best*. Ecosystems are an example of successful sustainable development, maybe the only ones we know. Genes store and reproduce the results of four billions years of biological evolution. In the ecosystem there are no parasites, nothing is useless, but everything has its own value. Natural ecosystems produce an enormous amount of biomass without creating any waste. Why do we prefer to replace nature through imperfect artefacts, instead of learning from that?
- *The awareness of limits to growth*. An "unlimited" economic growth combined with an exploding world population will lead to a societal collapse.[49] For a "good life"[50] the society needs today more decrease than growth; a fair redistribution of resources instead of an increasing inequality.[51] The actual decreasing birth rate in countries like Germany should be considered a huge opportunity – and not a huge curse. The world population of 2.5 billion in 1950 would offer much better possibilities to combine sustainability and prosperity than the expected population of 9 billion in 2040.
- *An "environmental" definition of responsibility and freedom*. The possibility to consume drugs, to spend a holiday in Miami or to choose among many products cannot really replace the real freedom of disobeying an unsustainable and sometimes

senseless social system; the freedom of self-realization has wider horizons. And that freedom is the real responsibility, because responsibility is neither obedience to the hierarchy nor a sense of duty in regard to the own company.

Because a "monoculture of sustainability" is a contradiction in terms, it is more correct to speak of different *cultures* of sustainability. These must not necessarily be invented. We can learn a lot from existing and past cultures, which were terminated by the colonisation or are endangered today by *the Westernization of the World*.[52]

The West can learn from its own history too. Two World Wars, Hiroshima, Chernobyl, financial crisis, climate change, etc. have turned this civilization into a rather low point of human development. According to the critic of modernity by Zygmunt Bauman, "first the rational determined world of the modern civilisation makes the holocaust possible".[53] A *collective memory*[54] is a good strategy against the repetition of such failures.

Christian Religion and Culture of Sustainability

"The Historical Roots of Our Ecological Crisis" was the title of an article published in 1967 by Lynn White, Professor of History at the University of California. His central thesis: "By destroying pagan animism, Christianity made it possible to exploit nature in a mood of indifference to the feelings of natural objects."[55] What caused the "huge burden of guilt" that Christianity bears?[56]

For White the modern science is "an extrapolation of natural theology". The Jewish-Christian tradition is not only responsible for a dualist view of the world, but also for a negative denotation of a part of the creation (Earth, nature, animals and body) in comparison to the otherworldly one (Heaven, human being and soul). In the same way the "mechanistic sciences" consider the soulless part of creation as a machine.[57]

With his program on "mastery over nature"[58] Francis Bacon wanted to realise what was written in the book of Genesis (1:28). God's appeal "Be fertile, increase in number, fill the earth, and be its master. Rule the fish in the sea, the birds in the sky, and all the animals that crawl on the earth" has never been more successful than in the last century.

Lynn White appealed for *an alternative Christian view*: "I personally doubt that disastrous ecologic backlash can be avoided simply by applying to our problems more science and more technology [...]: Despite Copernicus, all the cosmos rotates around our little globe; despite Darwin, we are not, in our hearts, part of the natural process [...] What we do about ecology depends on our understandings of the man-nature relationship. More science and more technology are not going to get us out of the present ecological crisis until we find a new religion, or rethink our old one [...] Possibly we should ponder the greatest radical in Christian history since Christ: Saint Francis of Assisi."[59] The "Canticle of the Sun" and the "Canticle of the Creatures" by

Francis of Assisi is a lesson of appreciation of the environment. White's appeal was the beginning of a new debate and orientation that is called *ecotheology*.

A sustainable influence of religion in general on society could be taken from the theories of the French sociologist Emile Durkheim. He described religion as "cement", which holds a society together in spite of the progressive division of labour and of utilitarianism.[60] If this cement becomes weaker (anomie), then social inequality and conflicts increase, as well as depression and the suicide rate.[61] At this point it is appropriate to ask in what way the unsustainability is a form of anomie, caused by a weakening solidarity and a parcelling of knowledge – and if a strengthening of the religious cohesion can promote a more sustainable development.

Unfortunately, Christian solidarity is not always unconditional. The subordinated role of the woman, the strong sexual moral and the difficultly dealing with different opinions within the Catholic church show, how community sense can force conformity instead of solidarity.

The economic politic of several "Christian parties" seems to confirm the thesis of Max Weber that the spirit of Capitalism has religious roots.[62] In the Middle Age God's indulgence could be "bought": the power accumulation increased the possibility to save ones own soul. Pope Leo X offered in 1517 indulgences for those who gave alms to rebuild St. Peter's Basilica in Rome. Martin Luther protested against this practice:[63] indulgence does not need any mediator and is an inner question between God and individual. Today we should ask: can the social and ecologic sins behind capital accumulation really be "excused" by prayer, charitable donations or sponsorships?

Dynamics

A central thesis of the previous chapter was "another development needs another culture". But how can we change the structures of our societal system and the dominant way of thinking? A *socio-cultural strategy for sustainability* is needed.

Elements of a socio-cultural strategy for sustainability

Ends do not justify means. The concrete result of a social and cultural change is the consequence of a strategy and of a process too – and not only of planned aims.

Every strategy for social change is designed according to a worldview that should be critically examined. Michael F. Jischa[64] classifies the most eminent strategies for sustainability (Efficiency, Consistency and Sufficiency)[65] in a matrix of different understandings of nature and social justice.

The strategies for sustainability should be examined against the background of the *understanding of humanity* too. In the Western culture at least two schools of thought could be distinguished:

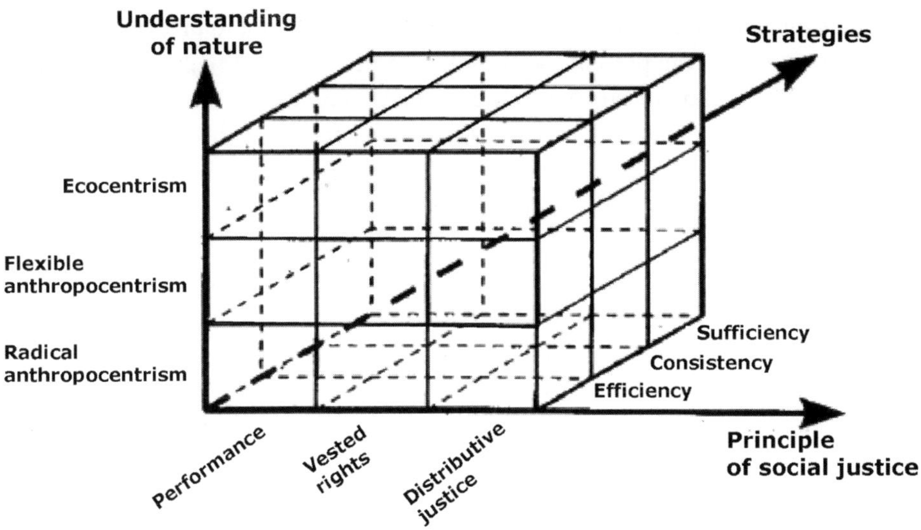

4 Matrix of Sustainability by Michael F. Jischa (1999)

- The first one is oriented to the negative understanding of humanity by the English philosopher Thomas Hobbes. Because the human being has an evil nature, we need laws and contracts, whose respect is guaranteed through the authority of the State.[66] Hobbes' philosophy was influenced by the cruelties of the Thirty Years' War that ended in 1648 with the Peace of Westphalia, which can be considered the first international peace treaty.
- The second one is based on a positive understanding of humanity. Immanuel Kant pleads the cause of culture not as civilization superstructure, but as reflection of an inner morality, that can be promoted through an organic education. Jean-Jacques Rousseau goes beyond that: human nature is good, but the constraints imposed by society and morals hinder the self-reliance: this repression makes humans into bad people. For Rousseau to free the human being from these "cultural chains" is the precondition for a peaceful coexistence. The French philosopher is a forerunner not only of the "culture critique", but also of psychoanalysis:[67] Freud's thesis of the human perversion as a consequence of an over-moral *Super-Ego* is not so far from Rousseau.

Several indicators are that our society is dominated by a negative understanding of humanity. Individualism, free competition, private property and privatisation could be expressions of the motto "one can really trust only oneself". We have to sign written contracts, because the spoken word isn't trustworthy enough. Army, police and bureaucracy are the result of an institutionalized mistrust of humanity.

The negative understanding of humanity is an important engine for scientific and technological progress. Because the human being is limited and imperfect, its capacities are supplemented by technologies[68] or substituted by more precise computers. "Social engineering" can prevent social conflicts and improve the collective efficiency. "Mistakes" or "unpleasant surprises" are avoided by automating social processes.

The debate on a strategy for sustainability is not free from this worldview, as the history of the German notion of *Nachhaltigkeit*[69] shows. In German forestry of the 18th century "Nachhaltigkeit" meant to harvest no more trees than can grow.[70] Later this principle was integrated by the aim of yield maximization.[71] The "sustainable forest management" has been able to guarantee that today 31 percent of the German territory is still covered by forest. But this forest is far from its original natural state. It is more an artificial and utilitarian construct, dominated by a few profitable species of trees, singly catalogued by age. The low biodiversity makes the forest prone to diseases and damages: actually only 31 percent of the German forest are completely healthy.[72]

Nevertheless some strategies for sustainability are oriented to the logic of the "sustainable forest management": they understand sustainability as additive *rationalization, optimization* and *modernization* of the society through knowledge, technology and norms. While the ideal head of such a strategy consists of experts, scientists, engineers and responsible managers (it reminds a little of Plato's Republic lead by philosophers), the democracy is a necessary evil, that guarantees legitimization and a *supporting participation* to a *top-down strategy*.

The socio-cultural strategy is based on a different understanding of sustainability. The main differences concern the following points:

- While the socio-cultural strategy aims for systemic and structural changes (because the causes of the global crisis are systemic and structural), the strategy for "sustainable growth" or of "carbon trading" aim for protecting the given social order through corrections.[73]
- Rationalization often happens through quantification and determination: while economic and technological aspects are centralized, social and cultural factors are instrumentalized, subordinated or even faded out. In "eco-modernization"[74] the same rationality that caused the crisis becomes the leading light of sustainability. At the same time cultural differences and traditions are considered disruptive factors that should be removed.[75]

  Instead socio-cultural strategies start from the assumption, that the main causes of the global crisis have a social and cultural origin – and should be counteracted on their own terrain. Social justice can be a strategy for the solution of ecologic and economic problems (e.g. sharing instead of possessing). Emotions are existential bridges to the external and inner nature of the humanity, whereas traditions can be proven strategies for survival in a specific environment.
- Every kind of modernization means a substitution of nature through artificiality – and does not really consider biophysical and human limits. Human life becomes

part of an efficient machine, whose complexity is meanwhile so high that it occupies massive intellectual and material resources: After a hard day of work nobody really has time for managing environmental crises or developing alternatives.[76] This quasi-totalitarian machine radiates a diffuse sense of security: the actual economic crisis shows how false and dangerous that feeling can be.
The socio-cultural strategy for sustainability starts from an awareness of the biophysical and human limits – and it means first a relief of the social systems. The motto "less is more" leads to a revaluation of the immaterial aspects of life and makes surely more sense than working overtime for high military expenditures or an unequal and virtual economic growth.

- Globalization aims for a reduction of complexity in a space, that should be as wide as possible (deterritorialization),[77] and in a period of time, that should be as short as possible (acceleration). In wide spaces there is a gap between institutions and citizens: the "media democracy" replaces the participated one. Globalization virtualizes social communication and privatizes public spaces and goods. Is a "social-ecologic globalization" really possible and desirable?
  Social-cultural strategies support the formation of *self-sustainable regions*,[78] whose dimension accords more to human limits. Democracy as well as fair economy can be better realised in small communities, where people meet, know and *trust* each other. The accent of this strategy is on public spaces and on face-to-face communication. Regions promote a *re-territorialization* of the economy and a cultural identification with the own landscape. A regionalized economy is less dependent on transport and petroleum than a globalized one. Regional "mistakes" can be better absorbed than global ones. Every region can learn from the others: cultural and political exchange is a matter of priority in comparison to the economic one.
- Top-down strategies risk legitimating the same social asymmetry that contributes to unsustainable development, or even the necessity of eco-dictatorships. Instead socio-cultural strategies aim for a "real" democracy, not as legitimization of a central power through occasional elections, but as an organizational expression of a "natural" principle: that the individual capacity to comprehend and to govern complexity can be expanded through dialogue and cooperation with other people (and not necessarily through technology). The precondition of this idea is a positive and evolutionary understanding of humanity.
- The socio-cultural strategies assume a "failure of the centrality", plead for a rediscovery of the *creative marginality*[79] and look for *change agents* in the "peripheries". Change agents "mustn't be invented, because they already exist as contradictory energies within the dominating social system. They are a multi-universe of subjects, behaviours, associations, deviant practices, cultures, that should be 'freed' from their ecologic or economic niches and supported in building the networks of 'cosmopolitan localism'."[80] *Networks* can combine different aspects, like local

and global, diversity and commonality, autonomy and unity. Therefore they are an ideal organization form for new types of *transversal alliances* between civil society, sustainable economy and culture (as societal sector). The universities, the churches or the abandoned villages in East Germany can become important *laboratories for sustainability* (or *sustainability pioneers*), in which new lifestyles and economy models can be tested and developed.[81]

A fundamental aspect of the socio-cultural strategy for sustainability concerns communication and media. For Luhmann no "presumable objective fact", not even an ecologic catastrophe, has any social effect, "as long as man does not *communicate* on that".[82]

Every culture needs *media* to be socialized – and this assumption is valid for a culture of sustainability too. *Media* makes communication between people independent from their distance as well as symbols makes the communication of objects independent from their physical presence. Through books, newspapers and television we can also "perceive" realities outside of our cognitive limits.[83] Through post, telephone and emails we can send instructions in places, we cannot manage personally. Through money we can trade with people whom we do not know. In globalization the construction of reality is a medial one.[84] The globality is *mediated* to us through technologic media. The international institutions, market and companies are *medial phenomena*: it is their strength, but also their weakness.

In the national state, military and police were a fundamental instruments of social control and of power expansion. Fordism brought this hierarchical and efficient organisation form into the factories.[85] This kind of power was visible, graspable, but also attackable, for example by trade unions.

The social order of globalization is different, guaranteed by a *cultural industry complex*.[86] It spreads the same values, modes, *logos*[87] and architecture around the world; universities that teach the same neoclassical economy models; newspapers that publish the same press release from the same press agencies.[88] "The television as vehicle of cultural globalization transmits the ideal of lifestyle of an American middle class family."[89]

Today the social system defuses the danger of a structural change not only by excluding, but also by appropriating alternatives: successful subcultures become commercialized or institutionalized;[90] public relations departments adapt the notion of sustainability to their own aims. This kind of power is diffuse, invisible and ungraspable, but very effective and difficult to attack.

Fortunately it has also some weaknesses that serve the socio-cultural strategy for sustainability. For example:

- A medial phenomena is somewhere also a virtual one. The financial markets show how powerful virtuality can be.[91] The actual financial crisis shows how fragile and replaceable virtuality can be.

- The existence of media depends on the presence of transmitters and receivers – and humans are fortunately not as controllable as technologies. In the long communication chain of globalization outputs seldom corresponds completely to inputs.[92] The contest of the transmitter is often different from the contest of the receiver. For people, who lose their jobs, the contingency of the economic crisis is much stronger than the relaxing message of TV-soaps.
- The medial power of globalization is reaching its limits, because it cannot really control the increasing social polarisation as resources run short. The world military expenditure reached in the last years new records.[93] Some western countries are going through an authoritarian phase.[94] The hope is that such developments shake up the minds.

Media are not only an instrument of social order. They can promote social change too – and it is the most important argument for a socio-cultural strategy for sustainability. The Bible is the best example of a medium that can change the world dramatically. A movie like "An Inconvenient Truth" by Al Gore made many people aware about climate dangers.

The cultural sector can be a fundamental engine of social change only provided that it is defunctionalized. A university, which is standardized and subordinated to the economic sector,[95] cannot really be a free laboratory for a better society. A press, whereas marketing plays a more important role than investigative journalism, cannot really support a democracy.

Herbert and Marshall McLuhan wrote *the medium is the message*:[96] what did they mean? The medium itself is transmitter and at the same time producer of cultural values. The globality is not only mediated, but probably also produced by media. If media contents are not independent from the media-type, then an oral culture will form a community in a different way than a written one. Watching the forest on TV is not experiencing the forest. A top-down-sustainability leads to a different development than a bottom-up sustainability. Therefore: for a social change oriented to sustainability we do not need only an "education for sustainability" or "TV for sustainability", but also *different media and communication structures*.

CHRISTIAN RELIGION AND SOCIO-CULTURAL STRATEGIES FOR SUSTAINABILITY

The history of the cooperation between political and religious power is long. Especially the organisational form of the Catholic church which is opposite to that of a democracy. In the past, disobeying the pope used to be punished by the Inquisition (a victim: Giordano Bruno); at present by excommunication (two victims: Leonardo Boff and Ernesto Cardenal).[97]

Among Christians there are very fundamentalist organisations (i.e. Opus Dei, the Society of St. Pius X, Christian Coalition) that support authoritarian political programmes.

The strong community sense of Catholics means also strong control of every member of the community: it inhibits the emotional and creative expression of individuals. In several religions arts, media and education are strongly functionalised.

Nonetheless, there are also several "sustainable" examples in the Christian religion. Liberation Theology promoted a bottom-up political change in Latin America, oriented towards social justice and solidarity. Christian organisations like Pax Christi, Misereor or Evangelischer Entwicklungsdienst[98] are integral parts of the international movement for *another world*. Churches, monasteries and missions are part of the last "public" places in a privatized society: they offer a possibility for reflexion and for personal communication.

Evolution

Biological evolution is the strategy that natural systems adopt to contrast the universal irreversible process of *entropy*[99] and to avoid the possibility of a systemic collapse. A not improbable danger: in the universe *Gaias* are exceptions.[100]

The application of theories of system to the sociological research implies the possibility of a *social evolution*. According to Jürgen Habermas every social change impetus can lead to an *innovation* or to a *dead-end*.[101] He defines "social evolution" as a "process of acquisition of learning mechanisms, that makes a society able to develop itself on different levels, so that it can escape the impending 'evolutive dead-ends' and become a 'good' society".[102] Social evolution means a "collective learning process and a societal improvement of learning capacities".[103]

There is a *cultural evolution* too, that interplays with the biological and social one (co-evolution). The gradual transformation of the ape into the human was primarily an intensive evolution of culture combined with the evolution of its genetic prerequisites (i.e. capacious brain, talking apparatus). While natural evolution operates through genes that store information and transmit it to the next generation, cultural evolution operates through non-physical replicators called *memes*.[104] The mixing of different genes strengthens the species against diseases, and the mixing of different memes expands the cultural spectrum of options to solve problems. If *monocultures* make ecosystems and social systems prone to crisis,[105] then *cultural diversity*[106] is for the sustainability of a society as important as the biodiversity for the crisis resistance of an ecosystem.[107]

What inhibits and what promotes a cultural evolution of our society?

Elements of cultural evolution

Inhibitory factors of cultural evolution can be found in unsustainable development. The characteristics of modernization and globalization – or simply of *The* development[108] – are the expression of a monological self-referentiality of the social system at the expense of its dialogical communication with the environment. In such a context the developmental dynamics aims for (a) a strong internal rationalization; (b) the internalization of resources from the environment; (c) the externalization of disorder, contradictions, conflicts and rubbish to the environment; (d) the transformation of societal boundaries into "dykes" and "walls", that protect the systems against environmental reactions (e.g. climate change, migrations, social conflicts) as well as against systemic changes (e.g. police against demonstrations, limited press freedom). These societal walls are designed according to *walls in the mind* – and strengthen walls in the mind by hindering environmental perception and communication in addition. The internalisation of these "walls" by the members of the social system causes alienation, isolation and conformity.[109]

The following graphic describes the development model of neoliberal globalization combining these considerations with the results of the analysis in the above paragraph on "complexity".

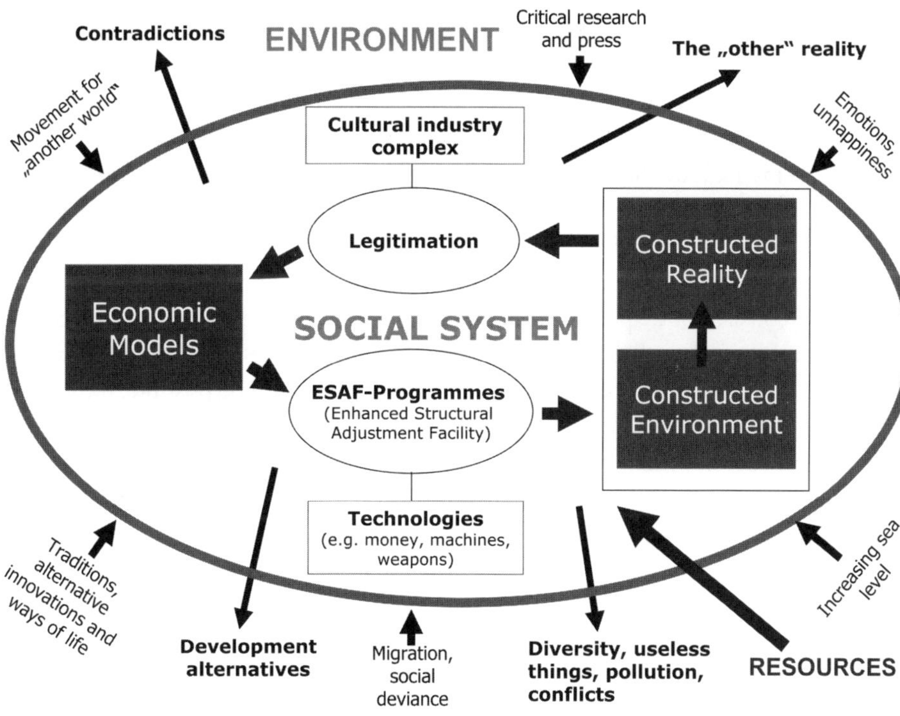

5 *The self-referential dynamics of the neoliberal globalization*

The environmental crisis is not only the result of a growing social and ecologic pressure on the boundaries of the social system, but also of an increasing *derealization*:

while the social system "feels" only the own artificial ordered reality, it is "insensitive" toward the rest.

This derealization is supported by the combined action of the following three factors:

- *Structures of social inequality*. Social inequality and power are reproduced and supported by social (e.g. networks, cliques), symbolic (i.e. status symbols) and cultural structures (i.e. selection in the educational system).[110] The inequality between nations, classes or people is the precondition of the possibility to internalize profits and externalize costs.[111] A context of social inequality promotes the universalization and institutionalization of the *strongest* argument – and not necessarily of the *best* one. In asymmetric social structures privileged people have more influence on the determination of development, whereas people who suffer more from development costs have no power, knowledge and capacity to change the situation. While poor people "think about bread and not about global problems",[112] rich people do not really worry about crises, because their wealth is an insurance for a personal good future. We should not forget: in the Cold War the bigger danger (besides the nuclear bombs) was the sense of security of those who possessed a nuclear bunker.
- *Technology*. Until technology can "adapt" the environment to the social system, the social system will avoid any adaptation to the environment. The "monetarization" of society offers the possibility to control anybody who needs money to survive. The mass media hide the reality through a diffusion of the viewpoint of the societal centres.[113]
- *Ideology (self-referential worldview)*. When the cause of problems are proposed as solution, then social systems are probably affected by ideology. Our dominant ideology consists of a specific understanding of nature, humanity and of course economy. Pierre Bourdieu analyses the neoliberal discourse as a form of "symbolic power": "the increasing economization of society would not be possible, if the market model did not advance to an ideal for all societal sectors and did not appear as 'natural' development without alternatives."[114] Fred Hirsch assumes that the (unequal) economic growth is not pushed by necessity, but by the culturally determined will to stand out from the mass.[115] In Western societies the question is not what kind of need, but which social status the need symbolizes. The "affiliation" to the upper-class is signed by the capacity to satisfy needs, that other classes cannot satisfy: what matters is not the absolute but the relative possession of money, of power or of knowledge, what one possesses more than others, the exclusivity. Economic growth (at the expense of nature and the Third World) is the strategy to guarantee social harmony in a democratic society in spite of an internal social inequality.[116]

Also the positive acceptance of technologies has today ideological traits. Part of the dominant ideology is the fear of conflicts, anarchy and chaos, which means the fear of any social change. Ideologies often produce, promote or use strong concepts of enemy to legitimate themselves.

The combined action of these three factors can be resumed with the motto: *Wisdom is power,*[117] *but power makes stupid*. Hierarchies and learnability tend to exclude each other: if one can impose his opinion on others, one must not learn from others. If one has been able to adapt the environment to one's own system of privileges, one will not be very interested to give up this system in order to protect the environment.

Unfortunately the social system cannot permanently lock out reality and environment – and sooner or later it pays a high cost for that.[118] A societal crisis is the consequence of a growing gap between worldview and reality. The reality that the social system excluded for a long time from their own perception, suddenly comes through the crisis into the system, massively.

What promotes a cultural evolution? An important difference between biological and cultural evolution is the human capability of *(self-)reflexivity*. On the one hand we can feel the coming dangers as well as our responsibility for destruction and suffering we cause. On the other hand (self-)reflexivity is also a chance to overcome the global crisis: the possibility to *learn a priori*. In the XXI century, sustainability cannot be a learning process *after* the "global collapse":[119] it is a sufficient motive for defining sustainability as the main *cultural* challenge in this century.

The circular self-reproductive mechanisms of the unsustainable development has to be broken through *cultural mutations*. Ideologies can be counteracted through a critical way of thinking, disobedience, intercultural exchanges, open political debates on contradictions. Of course arts are an ideal source of cultural mutations.

The sustainability of social and cultural systems depends from the *flexibility*[120] and the *sensitivity* of their boundaries. The *societal senses* percept the reality and politicize the inputs from the external and inner environment into the social system. Therefore the collective awareness of the impending global crisis and the necessity of a radical reaction depends on "healthy" societal senses. Who or what holds that function? For example non-governmental organisations, the civil society, an independent scientific research and a free press, maybe psycho-therapeutical groups. Migrants are an important bridge to the global reality, too.

Christian religion and cultural evolution

While cultural diversity is an important precondition of collective learning processes, the Christian religion is traditionally based on a strong assumption of exclusivity. Joseph Ratzinger considers the "dictatorship of relativism" as one of the worst evil of the Modern age.[121] With that assumption he goes the same missionary way that inspired colonization, the "Manifest Destiny" of the USA[122] as well as the globalization of a monoculture. The strong hierarchy of the Catholic church inhibits a necessary renewal of this confession. It's clergy often sticks to a conservative ideology, in that (their own?) sexual complexes get much more attention than the impending environmental catastrophe. Several scandals[123] seem to confirm the assumption of Sigmund Freud, that "hyper-morality" and "perversion" are often two sides of the same coin.[124] The strong sexual morality could have an inhibitory influence on social changes too.[125]

The superficial image of the Western as strong and courageous society covers up a deep fear of alternatives and of the alien. It could be a heritage of the atmosphere of fear in the Middle Ages too, when Christianity used a wide iconography of the devil for its propaganda. But the *Apocalypse*[126] is not a (divine) destiny, and another world is possible.[127] The "protestants" rebelled against the power of the clergy. In the Catholic church itself there are winds of change. The Christian *belief* in a better world promotes a cultural evolution. Christianity is a bridge between centres and peripheries and rich and poor people, but also between their own social role and their own humanity.

Conclusions

The story of the Easter Island reveals the *ambivalence* of (high) cultures: these people made their own island into a World Heritage Site,[128] simply by destroying themselves. But the Rapanui were not more "stupid" than modern humans. Despite two World Wars and the danger of a final Third World War, worldwide military expenditures reached in 2007 a record height of 1.34 trillions dollars.[129] While the worst ecologic catastrophe in millions of years is impending, governments throw away huge amounts of public resources to "save" the virtual financial markets and the automotive industry.

The Rapanui felled the last palm on their island while building the biggest Moai. In the same manner the "success" of the Western culture is today the major threat for our future.[130] The famous principle of the "survival of the fittest" by Herbert Spencer[131] evidently conflicts with another principle: the creature that wins against their own environment destroys itself.[132]

How can we avoid a "suicide of the fittest"?

The social and cultural dimensions of un/sustainability should be paid much more attention. A *cultural* perspective on societal development is indispensable, because:

- We cannot discern global problems through a microscope. Our parcelled knowledge needs an integrative moment that promotes an holistic worldview.
- In the natural, economic and social sciences there is a lack of *(self-)reflexivity*. The cultural perspective means an "ethnology of the culture to which we belong"[133] and a consideration of individuals as part of the observed reality, that they are

interpreting and changing at the same time.[134] There is no "neutral" scientific research.

- Although the western viewpoint is the dominant one (especially in the processes of globalization), it is not necessarily the best one or even the only one. The social sciences often describe the western society *as if* it were the whole world. The cultural perspective is not only a critical and undogmatic approach, but also a way to promote a necessary altercation with other viewpoints through an intra- and intercultural dialogue.
- The problems do not lead a society to the crisis more than the false reaction to the problems. The significance of cultural factors in a crisis situation is confirmed also by the fact that different societies react in a different way to similar challenges.[135]

What inhibits and what promotes a change of society into sustainability? I summarise some results from this chapter:

- Globalization is a "monocultural programme" for the global reduction of complexity and the standardization of diversities. Several *myths* promote a dangerous overestimation of human and technologic capacities. Today social inequality, mass media and technologies play a very important role in the social construction of reality and of the environment, that contributes to the impending global crisis. The sustainable development is a systemic and structural alternative to globalization, based on diverse cultur*es* of sustainability. The awareness of the human limits is the awareness of the relativity of our social and cultural production. Dialogue, cooperation, democracy, sharing of things and intercultural exchange belong to the "natural" strategies of supplementing the limited individual capacities through collective ones – but it needs a positive understanding of humanity.
- Also a culture of sustainability presupposes social providers and media to be effective. The "what" of a message is not independent from the "how" it is communicated. Sustainability does not mean more attention for environmental topics in the given university and TV, but in another university and TV.
- Societal crises are also caused by a growing derealisation of the social system – that is, by a strong self-referentiality, stiff boundaries and a weak environmental sensitivity. The combined action of social inequality, ideology and technology inhibits the social change into sustainability. However "collective learnability" is promoted by critical thinking, curiosity, imagination, intercultural communication and social laboratories. Civil society, (a defunctionalized) research, art and press are "societal senses" that promote the perception of reality and politicize environmental inputs to the social system. The combination of social system and personality makes an overall system able to evolve.[136] Cultural evolution is a fundamental component of sustainability. For sustainability the way is the goal and the goal is the way.

Because the structures of unsustainability are internalized by most people and institutions, "the capitalist society cannot be changed only through a will of radical change or through rational decisions".[137] The boundary between unsustainability and sustainability, between social order and alternatives goes through society, institutions and individuals.

Also Christianity is crossed by a strong *ambivalence*. On the one hand the Christian religion promotes an overestimation of the human being, on the other an awareness of its limits in front of God. On the one hand "religious beliefs" are out of question and exclude any real dialogue; on the other hand we need to "believe" in a reality outside of our cognitive limits, in not-yet-realities like social justice, peace or sustainability. Maybe God created the human in his own image, maybe humans created God(s) in their own image.[138]

It is very difficult to analyse how the ambivalence of Christianity is interconnected to the ambivalence of our society and culture, but both cases seem to prove what Michel Foucault said: no power can be possessed, but every kind of power has a relational character; there is no power without a countervailing power.[139]

In Christianity, in Western societies and in me there is a powerful order, but also dynamic movements. Although I belong to this society and culture (and they are part of mine), I wish for a radical change. The social change into sustainability is an interior evolution too,[140] that free our emotions and vital imagination, transforming the own alienation into cultural and political creativity.

NOTES

1 Diamond.

2 Carson.

3 The sociologist Bernd Hamm (1996) catalogues the indicators of the impending global crisis in its three dimensions. Growing world population, lack of resources, disappearance of biodiversity, cementification, climate change and health diseases (i.a.) are indicators of the ecological dimension. Public indebtedness, concentration of profit, deregulation, virtualization of the economy, unemployment (i.a.) belong to the economic dimension. Lack of democracy, corruption, social polarization, poverty, privatization of the public spaces, growing military expenditures and wars (i.a.) belong to the societal dimension. The cultural indicators (e.g. anomie, concentration on the press market) are listed under the societal category. Globalization leads to a stronger interplay and reciprocal alimentation of the three dimension of the global crisis.

4 Cf. Habermas (1979).

5 Diamond, 138.

6 Ibid, 118.

7 Ibid, 140.

8 Ibid.

9 Ibid, 121.

10 Ibid, 126.

11 Researchers founded a still bigger, but unfinished Moai in a quarry: 21 Meters in height and 270 tons in weight.
12 Diamond, 141.
13 Ibid, 142.
14 Ibid, 141.
15 Ibid, 147.
16 Socrates said "knowledge is virtue."
17 Boas, 589.
18 Cf. Kant.
19 Korzybski.
20 Cf. Bateson (1972).
21 Tolman, Downs, Stea.
22 Kandel.
23 Berger, Luckmann.
24 Hofstede and Hofstede.
25 Moebius, 97, on Michel Foucault's theories.
26 Wittgenstein.
27 Meadows.
28 Bourdieu (1982).
29 Esser, 375.
30 I.a. René Descartes (1596-1650), Francis Bacon (1561-1626) and Isaac Newton (1642-1726).
31 Speech of Winston Churchill in the House of Commons, 28 October 1943.
32 Horkheimer, 84-85.
33 Hösle, 59.
34 Peter Kareiva, 2008: "Ominous trends in nature recreation", PNAS 105, 2757-2758.
35 Jäger, Weinzierl, 70.
36 White.
37 "Sermon on the Mount", in: Cross.
38 Hösle, 17; Umweltrat der Bundesregierung (1994); Unesco (1998); Kurt, Wagner.
39 Tarozzi.
40 Kurt, Wagner, 13.
41 The most frequently quoted definition of sustainable development is that of the "Brundtland Report" (1987) by the World Commission on Environment and Development (WCED): "Sustainable development is a development that meets the needs of the present without compromising the ability of future generations to meet their own needs."
42 Cf. Eblinghaus, Stickler.
43 Cf. Tarozzi.
44 The main thesis of Dependencia theories is that the underdevelopment of countries and peoples is caused by the development of others: there is no poorness without richness. Therefore underdevelopment can be counteracted more by freeing from the exploitation (self-reliance) than through "development aid". This orientation, that had its most vivid period in the 60s and 70s in Latin America, represented the theoretical mouth of the meeting between Liberation Theology and Neo-Marxist analysis.
45 Commoner.

46 Vester.
47 Cf. Hösle.
48 Griefahn, 66.
49 Meadows (1972), (2004).
50 Nida-Rümelin, 7.
51 Hirsch.
52 Latouche.
53 Bauman, Ahrens, 27.
54 Halbwachs.
55 White.
56 Ibid.
57 Hösle, 54.
58 Bury, 48.
59 White.
60 Durkheim (1893), (1912).
61 Durkheim (1897).
62 Weber.
63 Article "Indulgences", in Cross. A famous quotation from Luther's Thesis 28 is: "As soon as a coin in the coffer rings, a soul from purgatory springs".
64 Former professor of mechanical engineering at the University of Clausthal and former President of the German branch of the Club of Rome.
65 The *strategy of efficiency* aims to solve ecologic problems through the development of new technologies and a more efficient use of resources. The *strategy of consistency* concerns the tolerability of anthropogenetic material flows for geogenetic material flows. The *strategies of sufficiency* plead a limitation of production and consumption as well as a demographic control.
66 Hobbes.
67 Nünning, 21.
68 McLuhan (1964).
69 The German translation of "sustainability".
70 Voss, 5.
71 Mantau, 1274.
72 German Federal Ministry for Agriculture (2009).
73 Michael S. Northcott's chapter in this book.
74 Jänicke.
75 Rieger, Leibfried, 13.
76 Ivan Illich defines the modern human as "homo miserabilis": because of unlimited needs, it is always poor, for example poor of time. Illich calculated, how much time an ordinary American spends for his car at the end of the 70s: for earning the money to buy it, for driving and for repairing it, he spends 1,600 hours (67 days) per year, 4.5 hours per day.
77 According to Alberto Magnaghi *deterritorialization* means for example, that everything can be produced or bought everywhere in every time; that cities or individuals are uprooted from their history and own identity.

78 According to Tarozzi and Magnaghi the fundamental principles of *self-sustainability* are: (a) Decentralization, self-reliance and self-government; (b) satisfaction of basic-needs; (c) eco-development and circulation's economy.
79 Guidicini, 12.
80 Magnaghi, 107.
81 Reisch, Scherhorn.
82 Luhmann (1992), 96-97.
83 According to Marshall McLuhan (1964) the media-technology is an extension of the human senses.
84 Cf. Merten, Schmidt and Weischenberg.
85 Sennett, 22-25.
86 Cf. Adorno, Horkheimer, 126-181.
87 Cf. Klein.
88 Hamm (2006), 280.
89 Bukow, Ottersbach, 73.
90 Two examples: organic food is today a "big business"; the environmental protest lead in the 80s to the establishment of Ministries for environment.
91 Brokers and investment bankers move everyday billions dollars around the world, as if they played a computer game. But the consequences are sometimes bitter reality for many other people. The distance between culprits and victims of the globalized economy is similar to the distance between bomber pilots and bomber victims.
92 Sennett, 32.
93 SIPRI.
94 E.g. in the USA, in Italy and in Austria.
95 Cf. the European Bologna Process.
96 McLuhan (1969).
97 Charismatic representatives of the Liberation Theology in Latin America.
98 <http://www.eed.de/en/>, 11 March 2009.
99 Rifkin.
100 Anne Primavesi's chapter in this book.
101 Habermas (2006).
102 Jäger and Weinzierl, 28.
103 Ibid, 30.
104 Dawkins.
105 Shiva.
106 Cultural diversity not only exists in the difference between Muslims, Jewish people and Christians, but also in sub-cultural diversity within the society.
107 Unesco (2001).
108 Sachs.
109 Chabot, 805-808.
110 Bourdieu (1983).
111 Chomsky.
112 Meadows, 13.
113 Cf. Merten, Schmidt, Weichsenberg.
114 Moebius, 133.

[115] Hirsch.
[116] Hösle, 32.
[117] Bacon, *Meditationes sacrae*, 11th article "De Haeresibus".
[118] Cf. the fourth principle of ecology by Barry Commoner: *There is no such thing as a free lunch.*
[119] Diamond, Beck.
[120] Finke, 264.
[121] Joseph Ratzinger, (2005, April 18th): *Missa pro Eligendo Romano Pontefice*, <http://www.vatican.va/gpII/documents/homily-pro-eligendo-pontifice_20050418_it.html>, accessed on 7 February 2009.
[122] It represents the American doctrine of the 19th century, that the USA has got a divine commission of expansion in the world. The Manifest Destiny was published by John L. O'Sullivan in 1839 in the journal "The United States Democratic Review" (cf. Stephanson).
[123] E.g. in Churches and Seminars in the USA, Switzerland, Austria and Germany.
[124] Freud.
[125] Reich (1949), (1963).
[126] Stefan Skrimshire's chapter in this book.
[127] Motto of the World Social Forum, <http://www.forumsocialmundial.org.br>, 23 May 2009.
[128] Unesco, <http://whc.unesco.org/en/list/715>, 5 March 2009.
[129] Cf. SIPRI (2008).
[130] Bovet, Rekacewicz et al., 46.
[131] Spencer, Vol. I, § 164.
[132] Bateson (1972), 514.
[133] Foucault, Carrette, 91.
[134] Cf. Bateson (1979).
[135] Diamond, 29.
[136] Habermas (2006).
[137] Bourdieu (1982).
[138] Feuerbach (1851), 241.
[139] Moebius (2009), 94.
[140] Bateson (1979), 199.

REFERENCES

Adorno, Theodor W. and Horkheimer, Max (1947), *La dialettica dell' illuminismo,* Torino: Einaudi (new edition 1966).

Bacon, Francis (1597), *Essayes: Religious Meditations, Plaies of Perswasion and Disswasion,* London.

Bateson, Gregory (1972), *Steps to an Ecology of Mind: Collected Essays in Anthropology, Psychiatry, Evolution and Epistemology*, Chicago: University of Chicago Press (new edition 2000).

— (1979), *Mind and Nature, A Necessary Unity (Advances in Systems Theory, Complexity, and the Human Sciences)*, New York: Hampton Press.

Bauman, Zygmunt and Ahrens, Uwe (1992), *Dialektik und Ordnung: Die Moderne und der Holocaust*, Hamburg: Europäische VA.
Beck, Ulrich (2007), *Weltrisikogesellschaft: Auf der Suche nach der verlorenen Sicherheit*, Frankfurt am Main: Suhrkamp.
Berger, Peter and Luckmann, Thomas (1966), *The Social Construction of Reality: A Treatise in the Sociology of Knowledge*, New York: Anchor Books.
Boas, Franz (1887), "Museums of Ethnology and their classification", *Science* 9, 485-486.
Bourdieu, Pierre (1982), *Die feinen Unterschiede: Kritik der gesellschaftlichen Urteilskraft*, Frankfurt am Main: Suhrkamp.
— (1983), "Ökonomisches Kapital, kulturelles Kapital, soziales Kapital", in: Reinhard Kreckel (ed.), *Soziale Ungleichheiten*, Göttingen: Schwartz, 183-198.
Brocchi, Davide (2008), "The Cultural Dimension of Sustainability", in: Sacha Kagan and Volker Kirchberg (eds.), *Sustainability: A New Frontier for the Arts and Cultures*, Frankfurt: VAS, 26-58.
Bukow, Dietrich and Ottersbach, Markus (1999), *Die Zivilgesellschaft in der Zerreißprobe*, Opladen: Leske + Budrich.
Bury, John (1932), *Storia dell'idea del progresso,* Milano: Feltrinelli (new edition 1979).
Carson, Rachel (1962), *Silent Spring*, New York: Houghton Mifflin.
Chabot, Sean (2008), "Love and Revolution", in: *Critical Sociology* 34, 803-828.
Chomsky, Noam (1998), *Profit Over People: Neoliberalism and Global Order*, New York: Seven Stories Press.
Commoner, Barry (1971), *The Closing Circle: Nature, Man, and Technology*, New York: Random House Inc.
Conti, Laura (1983), *Questo pianeta*, (This Planet), Rome: Editori Riuniti.
Cross, F. L. and Livingstone, E. A. (eds.) (2005), *The Oxford Dictionary of the Christian Church,*New York: Oxford University Press.
Darwin, Charles (1859), *On the Origin of Species by means of natural selection, or the preservation of favoured races in the struggle for life*, London: John Murray.
Dawkins, Richard (1976), *The Selfish Gene*, Oxford: OUP.
Diamond, Jared (2006), *Kollaps: Warum Gesellschaften überleben oder untergehen,*Frankfurt am Main: Fischer.
Downs, Roger and Stea, David (1973), *Image and Environment: Cognitive Mapping and Spatial Behavior*, Edison: Transaction (new edition 2005).
Durkheim, Emile (1893), *De la division du travail social*, Paris: Les Presses universitaires de France (new edition 1967).
— (1897), *Suicide*, New York: Free Press (new edition 1997).
— (1912), *Les formes élémentaires de la vie religieuse,* Paris: Les Presses universitaires de France (new edition 1968).
Eblinghaus, Helga and Stikler, Armin (1996), *Nachhaltigkeit und Macht: Zur Kritik von Sustainable Development*, Frankfurt am Main: IKO-Verlag.
Esser, Hartmut (2000), *Soziologie: Spezielle Grundlagen, Die Konstruktion der Gesellschaft*, Frankfurt am Main: Campus.
Feuerbach, Ludwig (1851), *Vorlesungen über das Wesen der Religion*, Leipzig: Verlag von Otto Wigand.
Finke, Peter (2003), "Kulturökologie", in: Nünning and Nünning, 248-279.

Foucault, Michel (1999), *Religion and Culture by Michel Foucault*, ed. by J. Carrette, Manchester: Manchester University Press.
Freud, Anna (ed.) (1986), *Sigmund Freud: The Essentials of Psycho-Analysis*, New York: Penguin.
German Federal Ministry for Agriculture (2009), *Ergebnisse der Waldzustandserhebung 2008*, Berlin.
Giddens, Anthony (ed.) (1972), *Emile Durkheim: Selected Writings*, London: Cambridge University Press.
— (1989), *Sociology*, Cambridge: Polity Press.
Griefahn, Monika (2002), "Nachhaltigkeitspolitik und Kulturpolitik – eine Verbindung mit Zukunft?" in: Kurt and Wagner, 59-67.
Guidicini, Paolo (1996), *Manuale per le ricerche sociali sul territorio,* (Handbook for Social Research on Territory), Milan: Franco Angeli.
Habermas, Jürgen (1979*), Communication and the Evolution of Society*, Boston: Beacon Press.
— (2006), *Theorie des kommunikativen Handelns*, Frankfurt am Main: Suhrkamp.
Halbwachs, Maurice (1992), *On Collective Memory*, Chicago: University of Chicago.
Hamm, Bernd (1996), *Struktur moderner Gesellschaften,* Opladen: Leske + Budrich.
— (2006), *Die soziale Struktur der Globalisierung,* Berlin: Kai Homilius Verlag.
Hirsch, Fred (1977), *Social Limits to Growth*, London: Routledge.
Hobbes, Thomas (1651), *Leviathan*, New York: Penguin (new edition 1982).
Hofstede, Geert, and Hofstede, Gert-Jan (2004), *Cultures and Organizations: Software of the Mind*, New York: McGraw-Hill.
Horkheimer, Max (1947), *Eclisse della ragione: Critica della ragione strumentale*, Torino: Einaudi (new edition 1969).
Hösle, Vittorio (1991), *Philosophie der ökologischen Krise*, München: Beck.
Illich, Ivan (1981), *Per una storia dei bisogni,* Milan: Mondadori.
Jäger, Wieland and Weinzierl, Ulrike (2007), *Moderne soziologische Theorien und sozialer Wandel,* Wiesbaden: VS Verlag für Sozialwissenschaften.
Jänicke, Martin (2007), *Megatrend Umweltinnovation: Zur ökologischen Modernisierung von Wirtschaft und Staat,* München: Oekom.
Jerman, Tina (ed.) (2001), *ZukunftsFormen – Kultur und Agenda 21,* Essen: Klartext.
Jischa, Michael (1999), "Das Leitbild Nachhaltigkeit und das Konzept Technikbewertung", *University of Potsdam: Brandenburgische Umweltberichte (BUB),* 5/1999, 81-94.
Kandel, Eric (2007), *Auf der Suche nach dem Gedächtnis: Die Entstehung einer neuen Wissenschaft des Geistes*, München: Pantheon.
Kant, Immanuel (1781), *Critique of Pure Reason*, New York: Dover Publications Inc (new edition 2004).
Klein, Naomi (2002), *No Logo!* 2. ed. Gütersloh: Riemann.
Korzybski, Alfred (1933), *Science and Sanity: An Introduction to Non-Aristotelian Systems and General Semantics*, Chicago: Institute of General Semantics, (new edition 1994).
Kurt, Hildegard and Wagner, Bernd (eds.) (2001), *Kultur – Kunst – Nachhaltigkeit,* Essen: Klartext Verlag.
Latouche, Serge (1996), *The Westernization of the World: The Significance, Scope and Limits of the Drive Towards Global Uniformity*, London: Paperback.
Lotman, Jurij M. and Uspenskij, Boris A. (1975), *Tipologia della cultura*, Milano: Bompiani.

Luhmann, Niklas (1984), *Soziale Systeme: Grundriss einer allgemeinen Theorie*, Frankfurt am Main.

— (1992), *Comunicazione ecologica: può la società moderna adattarsi alle minacce ecologiche?* Milan: Franco Angeli.

Magnaghi, Alberto (2000), *Il progetto locale*, Torino: Bollati Boringhieri.

Malthus, Thomas Robert (1798), *Essay on the Principle of Population*, New York: Dover Pubn. Inc (new edition 2007).

Mantau, Udo (1996), "Konstruktionsfehler im Nachhaltigkeitsdenken", *AFZ/Der Wald*, Nr. 23, 1274-1278.

McLuhan, Marshall (1964), *Understanding Media: The Extensions of Man*, Cambridge Mass.: The MIT Press (new edition 1994).

— (1969), *Das Medium ist die Botschaft*, Frankfurt am Main: Suhrkamp.

Meadows, Dennis (1972), *Die Grenzen des Wachstums*, Stuttgart: DVA.

— et al. (2004), *The Limits to Growth: The 30-Year Update*, White River Jct., Vermont: Chelsea Green.

Merten, Klaus, Schmidt, Siegfried and Weichsenberg, Siegfried (eds.) (1994), *Die Wirklichkeit der Medien: Eine Einführung in die Kommunikationswissenschaft*, Opladen: Westdeutscher Verlag.

Moebius, Stephan (2009), *Kultur*, Bielefeld: Transkript.

Nida-Rümelin, Julian (2001), "Partizipation im Kulturbetrieb", in: Jerman, 7.

Nünning, Ansar and Nünning, Vera (2003), *Konzepte der Kulturwissenschaften*, Stuttgart: J. B. Metzler.

Reich, Wilhelm (1949), *Character-Analysis*, New York: Orgone Institute Press.

— (1963), *La rivoluzione sessuale*, Milan: Feltrinelli.

Reisch, Lucia and Scherhorn, Gerhard (1998), "Wie könnten nachhaltige Lebensstile aussehen? Auf der Suche nach dem ethischen Konsum", in: *Der Bürger im Staat,* Stuttgart: Landeszentrale für politische Bildung, Nr. 2.

Rieger, Elmar and Leibfried, Stephan (2004), *Kultur versus Globalisierung*, Frankfurt am Main: Suhrkamp.

Rifkin, Jeremy (1981), *Entropy: A New Worldview*, New York: Bantam Books.

Sachs, Wolfgang (ed.) (1999), *The Development Dictionary: A Guide to Knowledge as Power*, London: Zed Books.

Sennett, Richard (2007), *The Culture of the New Capitalism*, London: Yale University Press.

Shiva, Vandana (1993), *Monocultures of the Mind*, London: Zed Books.

Stockholm International Peace Research Institute (SIPRI), *Yearbook 2008*, Oxford University Press, (<http://yearbook2008.sipri.org>).

Spencer, Herbert (1864-1867), *A System of Synthetic Philosophy; The Principles of Biology*, London & Edinburgh: Williams and Norgate.

Stephanson, Anders (1995), *Manifest Destiny: American Expansionism and the Empire of Right*, New York: Hill and Wang.

Stockholm International Peace Research Institute (SIPRI), *SIPRI Yearbook 2008: Armaments, Disarmament and International Security*, Stockholm: SIPRI.

Tarozzi, Alberto (1990), *Visioni di uno sviluppo diverso*, Torino: Gruppo Abele.

Tolman, Edward C. (1948), "Cognitive Maps in Rats and Men", *Psychological Review* 55, 4, 189-208.

Umweltrat der Bundesregierung (1994), *Umweltgutachten 1994,* Stuttgart: SRU.

Unesco (1998), *The Power of Culture,* Stockholm: Unesco-Conference.

— (2001), *Universal Declaration on Cultural Diversity – Adopted by the 31st session of the UNESCO General Conference*, Paris: Unesco.

Vester, Frederic (2002), *Die Kunst vernetzt zu denken: Ideen and Werkzeuge für einen neuen Umgang mit Komplexität*, München: dtv.

Voss, Gerhard (1997), *Das Leitbild der nachhaltigen Entwicklung – Darstellung und Kritik*, Köln: Institut der Deutschen Wirtschaft.

Weber, Max (1904), *Die protestantische Ethik und der Geist des Kapitalismus*, Erftstadt: Area (new edition 2005).

White, Lynn (1967), *The Historical Roots of our Ecologic Crisis*, in: A.E. Lugo and S.C. Snedaker (eds.), *Readings on Ecological Systems: Their Function and Relation to Man*, New York: MSS Educational Publishing (new edition 1971), <http://aeoe.org/resources/spiritual/rootsofcrisis.pdf>, 5 May 2009.

Wittgenstein, Ludwig (1921), *Tractatus Logico-Philosophicus*, New York: Dover Publications Inc. (new edition 1998).

World Commission On Employment und World Commission on Environment and Development (1987), *Our Common Future*, Oxford: Oxford University.

Mapping a Moral Landscape of the IPCC

David Olsson Kronlid

Introduction

Public discourse indicates that researchers, churches, economists, lay people, climate experts, policymakers, artists and educators agree that climate change[1] involves serious moral challenges to communities and individuals over the world. Still, with some exceptions moral philosophers and ethicists have not paid attention to the moral dimension of mitigation, adaptation and vulnerability.[2] Thus, this chapter and this book are unusual voices in a slowly expanding field of crossdisciplinary climate change research with the aim of highlighting meanings of ethical aspects of climate change.[3]

The main purpose of this chapter is to identify and clarify the ethical content of current climate change discourse. This is accomplished by answering the question of what a climate change moral landscape would look like according to the Intergovernmental panel of climate change (IPCC)[4] from the point of view of the capabilities approach (TCA). Thus, the results reported in this chapter stems from an ethical[5] analysis of selected documents from the IPCC's fourth assessment report (AR4).[6] This reading (or mapping) of the IPCC is related to climate change philosophy, anthropology, theology and mobility research.

Following this introduction, the second part produces a theoretical and methodological framework for the analysis and a short introduction to the capabilities approach. The third and main section presents results of the analysis. The final part includes conclusions and a discussion of the results as well as comments on and suggestions for further research.

Ethical pluralism and methodological mapping

The analysis takes the view that climate change research and public discourse benefits from characterising the moral dimension of climate change from varying ethical perspectives. This is based on three observations. The first observation regards that anthropogenic climate change involves serious moral conundrums regarding human wellbeing.[7] Hence, in general, clarifications are of importance. The second observation deals with the idea that climate change is a so-called post-normal problem. That is, it is characterised by harbouring perhaps unsolvable epistemic uncertainties, potentially catastrophic societal and ecological consequences as well as ethical, political, economic and scientific conflicts of value.[8] These complexities increase due to climate change's intersected and multileveled nature, which means that conflicts of interests, values and epistemic uncertainties need to be dealt with by taking local and global contexts into consideration while acknowledging private, political, and business interests in facing the past, present and future. The third observation concerns the fact that policy makers, climate justice researchers and public discourse presently frame these moral conundrums in a dominant paradigm of distributive justice. This implies that our understanding of the moral challenges of climate change is seriously limited in scope, currency and profile. This implies that our possible political solutions are seriously limited as well.

Taking these observations seriously, in the face of a widespread decrease in human wellbeing, I suggest that it is a poor strategy to lock climate change ethics to a specific paradigm however reasonable such a paradigm might be on its own terms. Rather, it is important to introduce and use theories with various currencies, scopes and profiles in mapping moral landscapes of climate change. Methodologically, this is an analytic approach called ethical pluralism that regards ethical theories as "... intellectual frameworks that support the analysis and solution of particular moral problems".[9] In other words, we may use ethical theories or conceptual frameworks as temporary and selective maps that describe certain moral aspects of climate change. Thus, such an ethical analytic mapping of climate change means entering a process of limiting:

> ... the objects of significance, the questions of relevance, and the strategies or methods of relevance. In short, we "map" the morally relevant situation according to our ... ethical theory, in the same way as we for instance "map" the ... world in accordance with geological theories.[10]

This chapter introduces an alternative model of a climate change ethic – climate capabilities. This model follows ethical pluralism as a methodological approach. Hence, the moral challenges of climate change are interpreted from a specific theoretical perspective while refraining from claiming that this perspective takes normative or methodological precedence over other theoretical perspectives.

The empirical material and other's work on climate change ethics are used as reminders of our wayfinding through moral landscapes in the process of mapping a climate change ethic.[11] Thus, mapping is a process of systematic "re-enactment, in narrative gesture, of the experience of moving from place to place within a region".[12] This is an important aspect of what climate change ethics means; a systematic re-enactment of our experiences in various moral landscapes that reminds us of various aspects of what it might mean to be a moral agent.

What are the moral challenges of climate change? Do the impacts of climate change threaten certain values and how can we deal with these threats? This chapter submits that the ethicist can only present temporary answers to this question and that we need a diversity of answers to capture the innate complexities of this question and to develop proper policy responses. The capabilities approach is one such answer.

THE CAPABILITIES APPROACH

As a map, the capabilities approach guides us according to its specific coordinates and has boundaries outside of which it leaves us without guidance.[13] The capabilities approach draws attention to whether the IPCC addresses intrinsic values of human wellbeing or whether certain intrinsic values are left unnoticed.

Bearing in mind that this short presentation cannot do the capabilities approach justice, some of its core assumptions will be introduced. The capabilities approach is a theoretical instrument for identifying "real freedoms that people have for leading a valuable life".[14] Accordingly, it is developed as a basis for assessments of human wellbeing[15] including principles of social justice, which can be the motivation for institutional guaranties.[16]

The intuition that every individual human being is entitled to be and do what he or she has reason to be and do is the foundation of the capabilities approach. These beings and doings are codified in the concepts of capabilities and functionings.[17] Whereas capabilities are actual abilities (positive freedoms) to act and to be, functionings are enacted capabilities.[18] Accordingly, a functioning is an achievement, whereas a capability is the ability to achieve. This twin-concept captures what real opportunities you have regarding the life you (may) lead:[19]

> Functionings are "beings and doings", such as being nourished, being confident, being able to travel, or taking part in political decisions. The word is of Aristotelian origin and, like Aristotle, this approach claims, significantly, "functionings are constitutive of a person's being".[20]

Theoretically, the space of capabilities delimits intrinsically valuable elements (capabilities and functionings) from instrumental valuable elements of human wellbeing (e.g. material and immaterial resources). Methodologically, when we "map" the IPCC according to the capabilities approach, we will be able to identify which (if any) intrinsic values that this mixed discourse of climate change research and policy addresses.

Accordingly, it is possible to analyse the ethical content in specific texts and various changes in a person's intrinsic quality of life: from knowledge to relationships to employment opportunities and inner peace, to self-confidence and various activities made possible by literacy classes.[21]

The question of whether we should develop specific set lists of capabilities (work, play, friendship, health, etc.) is well debated within capabilities research, dividing it roughly in to one open approach and one set list approach. Amartya Sen advocates that we should refrain from defining generic capabilities and functionings[22] because it is the "combination of foundational analysis and pragmatic use that gives the capability approach its extensive reach".[23] Further, he advocates an open approach because:

> To insist on a "fixed forever" list of capabilities would deny the possibility of progress in social understanding, and also go against the productive role of public discussion, social agitation, and open debates.[24]

Nussbaum,[25] Alkire and Black[26] and Robeyns[27] develop generic and contextual set lists of capabilities arguing that an open approach lacks "critical bite when it comes to assessing norms"[28] and that this makes it at the risk of becoming incomplete[29] and vague.[30] Hence, these and other capability researchers argue for a specific set list (or specific set lists) of capabilities emphasising the normative dimension of the capabilities approach.[31]

The analysis of IPCC reported in this chapter occupies a middle ground between these two positions and utilizes specific methodological criteria[32] for using a *tentative* set list of generic climate capabilities as a map. A generic list is the proper abstraction level for mapping the moral landscape of IPCC since IPCC offers generic statements regarding climate change mitigation, adaptation, and vulnerability and resilience.

According to Alkire and Black, Finnis' list in table one of generic dimensions of wellbeing represents "the simplest, self-evident explanations for humanly worthwhile activities".[33] This involves excluding dimensions that are merely means to wellbeing and elements that is covered by any of the other dimensions.[34] Accordingly, the dimensions on this list[35] represent some of the most basic reasons for acting, reminisce of distinct positive freedoms[36] and are vantage points for the study of IPCC-documents reported in this chapter:

In the following, capabilities and functionings that are vulnerable (or predicted to be vulnerable) to climate change impacts are referred to as *climate capabilities*.

A LIMITED LANDSCAPE

Does or will climate change affect people's capabilities according to the IPCC? The answer to this question is that we do not know, because the IPCC does not address health, mobility, work, etc. as capabilities. However, the following scenery of climate capabilities emerges when travelling through the landscape of selected parts of AR4 with this particular capabilities-map in our hands.

AR4 addresses five themes that correspond to or indirectly relate to the set list of capabilities (table one): life, knowledge (but not the appreciation of beauty), work (but not play), transcendence, other species and mobility. The following discussion focuses on the capabilities of Knowledge and Appreciation of beauty, Work and Play and Transcendence and Mobility.[37]

Aim/scope of the list		*Generic Content* – Based on Alkire & Black, 1997; Robeyns, 2003; Nussbaum 2000; and Kronlid 2008b.
Dimensions	1. Life	1. Life itself – its maintenance and transition – health and safety.
	2. Knowledge and appreciation of beauty	2. Correlative to humans being rational and their capacity to "know reality and appreciate beauty".
	3. Work and play	3. Some degree of excellence in work and play. Correlative to humans being "simultaneously rational and animal" and their resultant capacity to "transform the natural world by using realities, beginning with their own bodily selves, to express meanings and serve purposes".
	4. Friendship	4. Coherence between and among individuals and groups of persons – living at peace with others, neighbourliness, friendship.
	5. Self-integration	5. Coherence between the different dimensions of the person, that is, "inner peace".
	6. Coherent self-determination	6. Practical Reasonableness, coherence among ones Judgements, choices and performances – peace of conscience.
	7. Transcendence	7. Religion, spirituality – coherence with some more-than-human-source of meaning and value
	8. Other species	8. Coherence with the life of animals, plants, and the world of nature
	9. Mobility	9. Correlative to humans being mobile in existential, social and geographical space.

Table 1: *Generic Capabilities*

KNOWLEDGE AND APPRECIATION OF BEAUTY

The question of knowledge concerns people's ability to practice theoretical and practical knowledge. In a capabilities-reading, it is important not to use a fixed understanding of knowledge. Hence, methodologically this capability represents any system of theoretical and practical knowledge that is rational and meaningful (coherent) on its own terms.

The material does not refer to appreciation of beauty or any aesthetic aspects related to work thus the following quote underlines loss or change of knowledge systems due to ecosystem shifts caused by climate change:

> The loss of local knowledge associated with thresholds in ecological systems is a limit to the effectiveness of adaptation.[38]

The particular knowledge systems addressed are related to environmental adaptation practices. It is argued that climate change causes shifts and flips in ecosystems that make the traditional adaptation practices inadequate. Hence, it is argued that indigenous knowledge systems (IKS) are threatened by climate change.

This claim is affirmed by indigenous people and by research about indigenous people's responses to climate change.[39] For example, Timothy B. Leduc argues that the changing climate modifies the adaptation knowledge of Inuit – *Inuit Qaujimajatuqangit* (IQ). One of his informants states:

> IQ was real at the time because the planet was not changing. They knew what was going to happen to the weather for four seasons. But today the weather has so changed that IQ is pretty much gone, it can no longer predict because of the change in climate.[40]

One of the World Wildlife Fund's climate witnesses Sámi reindeer herder Olav Mathis Eira from Norway states:

> During the last 20 years I have observed various changes in the climate. The most urgent change for us, the Sámi people who live off the reindeer, has been the winter rains. Rain in the winter is normally very rare this far North. In the old days this used to happen only every 30 years and we had ancient methods of foretelling the weather. Now this is no longer possible.[41]

The climate in flux also affects western or modern knowledge systems (MKS), which the IPCC terminology underlines.[42] The IPCC communicates uncertainties of consequences of present and future climate change in a terminology of very high, high, medium, low, and very low confidence (ranging from at least a 9 out of 10 chance to less than 1 out of 10 chance) of being correct. Furthermore, The IPCC communicates uncertainties in a terminology of likelihood regarding "probabilistic assessment of some well-defined outcome having occurred or occurring in the future"; virtually certain, very likely, likely, about as likely as not, unlikely, very unlikely, and exceptionally unlikely ranging from > 99% probability of occurrence down to < 1% probability.[43] This terminology reflects that climate change research, i.e. knowledge involves theoretical as well as empirical uncertainties and how MKS deals with these uncertainties.

This calls our attention to that both indigenous and modern knowledge systems are vulnerable to the post-normal nature of climate change. This positions indigenous knowledge and western knowledge on equal terms of reference. Thus, it highlights that experiencing risks of loosing certain abilities to be "rational" in the face of a changing climate can, despite differences in adaptation capacity, vulnerability and material standard, be equally threatening to indigenous and modern[44] cultures.[45]

WORK AND PLAY

The capability of work and play concerns some degree of excellence in work and play. Since The IPCC does not mention play, this discussion will focus on the ability to work. The citation below reminds us of how resource dependent work among the poor and among women is and will be affected by climate change:

> There are structural differences between men and women through, for example, gender-specific roles in society, work and domestic life. [...] As resource-dependent activities are directly dependent on climatic conditions, changes in climate variability projected for future climates are likely to affect women through a variety of mechanisms: directly through water availability, vegetation and fuel wood availability and through health issues relating to vulnerable populations (especially dependent children and elderly). Most fundamentally, the vulnerability of women in agricultural economies is affected by their relative insecurity of access and rights over resources and sources of wealth such as agricultural land. It is well established that women are disadvantaged in terms of property rights and security of tenure, though the mechanisms and exact form of the insecurity are contested (Agarwal, 2003; Jackson, 2003). This insecurity can have implications both for their vulnerability in a changing climate, and also their capacity to adapt productive livelihoods to a changing climate.[46]

Furthermore, The IPCC underlines work changes in European communities dependent upon ski tourism. Such changes include reductions in working opportunities and changed working conditions regarding "[a]rtificial snow-making; grooming of ski slopes; moving ski areas to higher altitudes and glaciers; use of white plastic sheets as protection against glacier melt; diversification of tourism revenues (e.g., all-year tourism)".[47] In addition, erosion of beaches and coral bleaching affecting local resources in small islands will have an impact on e.g. small fisheries, which will also affect people in the tourism industry.[48] Moreover, people's ability to execute excellence in their work in agriculture, forestry and industry, settlement and society sectors will be affected by extreme weather and climate events in the mid- to late 21st century (without taking changes or developments in adaptive capacity into account).[49]

For example, increased heavy precipitation is very likely to damage crops, increase soil erosion and reduce the ability to cultivate land. Additionally, flooding caused by increased precipitation will very likely cause disruption to settlements, commerce and transport and put pressure on urban and rural infrastructures. Moreover, increased land degradation, lower yields/crop damage and failure and increased livestock deaths as well as increased risk of food and water shortages with reduced hydropower generation potentials and the potential for population migration are likely in areas affected by drought.[50]

The IPCC predominantly addresses work as a resource and instrumental to human wellbeing. From a capabilities approach however, climate change directly threatens human wellbeing as such. Thus, a capabilities-reading of the IPCC's results and predictions indicates that it is very likely that climate change threatens people's abilities to execute excellence in their work in affluent as well as scarce communities.

Transcendence

The ability to transcend refers to a person's ability to relate to some more-than-human-source of meaning and value. The IPCC highlights statistical relationship between adaptation and religion:

> Empirical research has shown that entitlements to elements of adaptive capacity are socially differentiated along the lines of age, ethnicity, class, religion and gender (Cutter, 1995; Denton, 2002; Enarson, 2002).[51]

However, there is no reference to the opposite relationship, i.e. that climate change affects people's ability to transcend. What this might mean will now be further elaborated.

Here, transcendence is analogous to religious praxis, which has cognitive and ideological, social-ethical, ritual, institutional, aesthetic and psychological contextual dimensions.[52] Thus, varying trajectories in people's lives situated in specific socioeconomic, ecologic and cultural contexts fabricate the ability to transcend.[53] Furthermore:

> In a wider perspective, religion can comprise individual and social views, ideologies, values and beliefs, without any necessary relation to a given religious institution or tradition [and in] a more restricted sense, religion comprises only that which believers themselves explicitly distinguish from more worldly phenomena.[54]

This means that a person's ability to transcend is interrelated with other important everyday faculties of life. Leduc's crossdisciplinary research on indigenous transcendence practice provides us with an example of how specific indigenous transcendence is affected and threatened by climate change.

Leduc's work focuses on problems associated with how climate change research – in its effort to understand and make use of Inuit apprehensions of a changing environment – gets involved in reductionist interpretations of Inuit spiritual conceptions. In dealing with this problem, he focuses on the Inuktitut term "Sila" that has been interpreted as an Inuit spiritual power associated with "weather" in climate change research.

Leduc reaches two meanings of Sila through a triangulating transdisciplinary research approach[55] that involves an ongoing dialogue between him and Jaypeetee Arnakak, Inuit policy analyst from Igualit, Nunavut. The first meaning of Sila is Silarjuaq, which refers to something great that simply is and is the substance of life. The second meaning of Sila is Silatuniq, which refers to "a practical wisdom for living within this land"[56] that can attend to Silarjuaq.[57]

Leduc concludes that the interrelated doings (Silatuniq) and beings (Silarjuaq) of the whole of Sila are vulnerable to climate change in two important ways. First, the Inuit ability to transcend is being threatened by the ways in which climate change researchers anthropomorphise and reduce its meaning to fit modern environmental management models. Second, it is threatened because the environmental pressure of the changing climate induces an enforced uncertainty regarding their understanding of Sila.[58] In short, climate change and the modern climate change discourse threaten Sila as a specific contextual transcending practice, which might transform the transcendence practice of this particular Inuit community.

Mobility

The IPCC underlines that climate change affects voluntary and involuntary geographical movements:

> In the Sudano-Sahel region of Africa, persistent below-average rainfall and recurrent droughts in the late 20th century have constricted physical and ecological limits by contributing to land degradation, diminished livelihood opportunities, food insecurity, *internal displacement* of people, *cross-border migrations* and civil strife (Mortimore and Adams, 2001; Leary et al., 2006; Osman-Elasha et al., 2006).[59]

The IPCC addresses mobility as both an adaptive strategy for people living in communities that reach their sustainability or resilience[60] threshold and as a limit of adaptation strategies. Regardless of whether displacements and cross-border migrations are part of reactive[61] adaptive strategies or represent lack of adaptive strategies, we need to understand how social and geographical mobility interlink in order to understand fully how climate change affects mobility.[62]

Margaret Grieco and Julian Hine relates "stranded mobility"[63] to how the hurricane Katrina affected people's wellbeing in their essay on the interaction of geographical mobility and social exclusion:

> The evidence is clear: those without access to private means of transport died, were displaced, were impoverished and were scattered beyond their control during Katrina and its aftermath.[64]

Here, mobility research highlights that constrained social and geographical mobility collude in times of crisis: "Routine neglect of the relationship between transport and social exclusion was the mother of the New Orleans crisis."[65] Thus, this suggests that the extent to which climate change threatens a person's geographical mobility is related to that person's ability to be socially mobile and that it is "simply nonsensical"[66] to separate geographical from social mobility.

The IPCC connects social mobility and geographical mobility: "the spatial patterns of existing social networks in a community influence their adaptation to climate change", because it determines "the success and patterns of migration as an adaptive strategy".[67]

Clearly, the IPCC regards mobility as a recourse for dealing with climate change impacts. This approach has its advantages. Conceptualising mobility as a recourse for wellbeing clarifies how stranded social/geographical movements due to socially induced unequal distribution of access to recourses (e.g. vehicles) may likely increase climate change induced suffering. Further, it contributes to (as in the case of Katrina) marking out the limits of social/geographical mobility as adaptation strategy in times of environmental crises.

The IPCC and mobility research share the idea that environmental crises, such as climate change, visualises the identities of the mobility poor and those with access to mobility resources, consequently delimiting climate change vulnerable from resilient individuals and social groups. Furthermore, access to resources for social mobility (such as e.g. access to social networks) sets limits for access to geographical mobility; if your social position does not allow for a well-established and sustainable social network – if you cannot decrease your vulnerability by moving – what is the point of leaving? Finally, this highlights that climate change induced stress on resource systems (private car ownership, factual public transport, etc.) in geographical space may further cement the boundary between the vulnerable and non-vulnerable in developing and developed countries.

All work and no play

The IPCC does not address life, knowledge, work, religion, mobility, etc. *as* capabilities. Moreover, several of the items on the list in table one are missing from the IPCC document study reported in this chapter. Still the IPCC confronts several topics, which correspond to the capabilities accounted for in table one and it is therefore possible to make a constructive interpretation of the result of this mapping – it reveals the following.

First, the topography of the moral landscape that emerges is quite limited as the leading body for the assessment of climate change at best brushes past the highs and lows of human existence. The IPCC leaves abilities key to human wellbeing and dignified life – appreciation of beauty, play, friendship, self-integration, coherent self-determination – unmentioned. As we are facing a deeply uncertain future where climate change increases pressure on human flourishing, climate change research and policy thus needs to include systematic analyses of how climate change affects such key dimensions of flourishing such as play, sex, friendship, creative aesthetics and learning.

Second, the capabilities approach is a promising analytic framework for mapping the moral landscape of the IPCC. It highlights that climate change threatens not merely valuable resources and people's welfare but also abilities intrinsic to human wellbeing. Hence, climate change justice goes beyond fair distribution of adaptation resources, mitigation responsibilities and the burdens and benefits of climate change effects.

Third, it becomes obvious that the abilities to know, work, transcend and be mobile are intimately connected.[68] For example, the ability of theoretical and practical knowledge, of work such as reindeer herding, of transcending as concrete spiritual being and doing are deeply interrelated in the Inuit example above. Furthermore, that these abilities are individual/social faculties and that social and geographical mobility are interrelated strengthens this connection.

Fourth, from this follows that if climate change threatens one or several of these capabilities it is likely that other capabilities will face increased vulnerability as well. If this is true, strengthening i.e. the ability to transcend may strengthen one, several or all the other capabilities as well. This result is of importance for present and future climate change policy, international climate change negotiations, education for sustainable development and the development of individual and community adaptation practices.

Fifth, a capabilities-mapping of the IPCC highlights the importance of looking at conversion factors and the lack thereof. As Robeyns puts it "if a person is disabled, or in a bad physical condition, or has never learned to cycle, then the bicycle will be of limited help to enable the functioning of mobility."[69] For example, being able to be mobile includes being able to "convert the characteristics of the commodity [e.g. a vehicle] into a functioning".[70] On a personal level, factors such as "metabolism, physical condition, sex, reading skills, intelligence" may have prevented the mobility poor of New Orleans to be sufficiently mobile. Furthermore, following Cresswell the politics of mobility involves social factors like policies, norms, discriminating institutional practises and discourses, roles associated with gender, ethnicity, age, societal hierarchies and power relations. Thus, a capabilities-reading of how climate change threatens certain of our individual capabilities opens a space for systematic critical reflections on the role of personal, social (including political) and environmental conversion factors in such processes.

Finally, this mapping of the IPCC reveals the limit of climate change science because it shows how specific faculties of life, which are intrinsic to human wellbeing, are systematically excluded from the most authoritative body of climate change research. This underlines the point that climate change research needs to include and reinforce crossdisciplinary research, research from the humanities and other ethical paradigms than the intergenerational distributive justice model.

Notes

1 The term "climate change" refers to anthropogenic changes in the climate throughout this chapter.

2 Page, Northcott, Garvey and Adger et al. are some important exceptions. See also Ott in this volume.

3 This chapter is part of the research project *Climate Capabilities* (2008-2010) financed by the Swedish research council FORMAS.

4 IPCC is the leading body for the assessment of climate change. Chaired by Mr. Rajendra Pachauri, IPCC synthesises current climate change research in their assessment reports.

5 Ethical analysis or "ethics" means "the philosophical inquiry into the nature, extent and justification of the ethical claims which are made on human beings ... " (Dower, 2-3).

6 The analysis focuses on the Working group II's (WGII) *the Summary for Policymakers* (SPM) and chapter 17, *Assessment of adaptation practices, options, constraints and capacity* (Ch. 17). The summary for policymakers contains the key scientific findings and is the core document in which IPCC communicates its results to media, academia and policymakers. Hence, from a discourse theoretical point of view SPM is the most authoritative among the IPCC documents.

7 The fact that climate change affects people's health, mobility, education, work, consumption, etc. is clearly an indicator of this.

8 Bäckstrand, 94. My translation from Swedish.

9 Stone, 133.

10 Kronlid (2003), 55-46.

11 Dower, 2-3. "Ethic" is the result of ethics, e.g. sets of principles, conceptual frameworks or full-blown theories).

12 Ingold, 232.

13 One important difference between modern geographical maps and conceptual framework maps is of course the latter's often strong ambition to map several different landscapes with only one map, whilst geographical maps are considered to correspond only to one location.

14 Robeyns (2003), 61.

15 Nussbaum (2005), 6.

16 Nussbaum (2002), 98.

17 Alkire and Black, 268 "prefer the term 'flourishing' – which also communicates the sense that people pursue and participate in but never fully realize" the dimensions of wellbeing "once and for all". I use "functions" and "flourishing", and "elements on capability set lists" and "dimensions of flourishing" interchangeably.

18 Alkire, 117. See also Sen (1993), 31: "functionings represent parts of a state of a person – in particular the various things he or she manages to be and do in life. The *capability* of a person reflects the alternative combinations of functionings the person can achieve, and from which he or she can choose one collection." From Qizilbash, 157.
19 Sen (1987), 36; Grasso, 240.
20 Alkire, 118.
21 Alkire, 119.
22 Sen is only critical of "one pre-determined canonical list of capabilities, chosen by theorists without any general social discussion or public reasoning" (Sen (2005), 158).
23 Sen (1999), here from Qizilbash, 159.
24 Sen (2005), 160.
25 Nussbaum (2003); (2005), 178-179.
26 Alkire and Black.
27 Robeyns (2003), 71-72.
28 Qizilbash, 159.
29 Nussbaum, (2005a), 13-14.
30 Nussbaum, (2003), 46-47.
31 Nussbaum, (2003), (2005a), (2005b), Qizilbash, 157.
32 See Robeyns, (2003).
33 Alkire and Black, 272.
34 Alkire and Black, 269, Kronlid (2008a).
35 Two dimensions are added to Finnis' list: Other species (Nussbaum (2005)) and Mobility (Robeyns, (2003) and Kronlid (2008a)).
36 Alkire and Black, 268.
37 A discussion of the missing capabilities (appreciation of beauty, play, friendship, self-integration, and coherent self-determination) and how they relate to other dimensions of wellbeing are of significant importance and are dealt with in coming publications.
38 Ch. 17, 734.
39 Research about the changing climate's influence on indigenous knowledge systems warrants this. See e.g. UNESCO (2009) and Bates.
40 Leduc (2007), 247.
41 <http://www.panda.org/about_our_earth/aboutcc/problems/people_at_risk/personal_stories/witness_stories/?uNewsID=113580>, 22 May 2009.
42 This point is also made by Leduc.
43 SPM, End Box 2, 21.
44 The term modern does not imply that indigenous cultures are pre-modern. Rather, "modern" refers to life styles and knowledge systems typical for affluent growth economies characterized by de-contextualised education, reductionist research and learning and a separation of the spiritual and the worldly.
45 Contextual and individual differences in vulnerability and power can of course produce various consequences of such experience.
46 Ch. 17, Box 17.5, 730.
47 Ch. 17, Table 17.1, 722.
48 SPM, 15.
49 Ch. 17, Table 17.1, 722.

50 Ch. 17, Table 17.1, 722.
51 Ch. 17, Box 17.5, 730.
52 Bergmann (2005), 56.
53 Bergmann (2003).
54 Bergmann (2005), 56.
55 Leduc, 241.
56 Leduc, 244.
57 Leduc, 243.
58 Leduc, 247-248.
59 Ch. 17, 734. My italics.
60 "Resilience" refers to "the self-repairing capacity of ecosystems" (Folke et al., 558) but could also refer to the self-repairing capacity of eco-social systems.
61 Adger, Paavola and Huq, 8. In this chapter, adaptation is understood as institutional (inter- and transnational, national, and local) and/or individual proactive, reactive and inactive responses to climate change.
62 Mobility researchers also allude to a dimension of mobility sometimes referred to as the meanings attached to geographical mobility, sometimes as existential mobility, and sometimes as symbolic mobility. See for example Cresswell and Kronlid (2008b).
63 "Stranded mobility is a concept developed in South Africa to refer to the situation of constrained mobility [movement] and poor accessibility experienced by the black townships." Grieco and Hine, 65.
64 Grieco and Hine, 67.
65 Grieco and Hine, 67. Se also Cresswell.
66 Cresswell, 261.
67 Ch. 17, 734.
68 Leduc argues that indigenous knowledge systems are developed in intimate relation to the land (this volume).
69 Robeyns (2005), 99.
70 Robeyns (2005), 99.

References

Adger, W. Neil, Paavola, Jouni, Saleemul, H. and Mace, M. C. (2006), *Fairness in Adaptation to Climate Change*, Cambridge/London: MIT Press.

Alkire, Sabina (2005), "Why the Capability Approach?" *Journal of Human Development*, Vol. 6, 115-133.

Alkire, Sabina and Black, Rufus (1997), "A Practical Reasoning Theory of Development Ethics: Furthering the Capabilities Approach", *Journal of International Development*, Vol. 9, 263-279.

Bates, Peter (2009), "Learning and Inuit Knowledge in Nunavut, Canada", in: UNESCO (2009).

Bergmann, Sigurd (2003), *God in Context: A Survey of Contextual Theology*, Aldershot: Ashgate.

— (2005), *Architecture, Aesth/Ethics & Religion*, Frankfurt am Main/London: IKO-Verlag für interkulturelle Kommunikation.

Bergmann, Sigurd, Hoff, Thomas and Sager, Tore (eds.) (2008), *Spaces of Mobility: Essays on the Planning, Ethics, Engineering and Religion of Human Motion*, London: Equinox.

Bergmann, Sigurd and Sager, Tore (eds.) (2008), *The Ethics of Mobilities: Rethinking Place, Exclusion, Freedom and Environment*, Aldershot: Ashgate.

Bäckstrand, Karin (2003), "Samspelet mellan vetenskap och politik: Experternas, medborgarnas och beslutsfattarnas roll i miljöforskningen och mljöpolitiken", in: Lundgren, 73-98.

Climate Change (2007): Impacts, Adaptation and Vulnerability, Contribution of Working Group II to the Fourth Assessment Report of the Intergovernmental Panel on Climate Change, M.L. Parry, O.F. Canziani, J.P. Palutikof, P.J. van der Linden and C.E. Hanson (eds.), Cambridge: Cambridge University Press, 976.

Cresswell, Timothy (2006), *On the Move: Mobility in the Modern Western World*, New York/London: Routledge.

Cresswell, Timothy and Priya Uteng, Tanu (eds.) (2008), *Gendered Mobilities*, Aldershot: Ashgate.

Dower, Nigel (1998), *World Ethics: The New Agenda*, Edinburgh: Edinburgh University Press.

Folke, Carl, Hahn, Thomas, Olsson, Per and Norberg, Jon (2005), "Adaptive Governance of Social Ecological Systems", *Annual Review of Ecology, Evolution, and Systematics*, Vol. 30, 441-473.

Folke, Carl, Carpenter, Steve, Walker, Brian, Scheffer, Marten, Elmqvist, Thomas, Gunderson, Lance and Holling, C.S. (2004), "Regime Shifts, Resilience, and Biodiversity in Ecosystem Management", *Annual Review of Ecology, Evolution, and Systematics*, Vol. 35, 557-581.

Garvey, James (2008), *The Ethics of Climate Change: Right and Wrong in a Warming World*, London/New York: Continuum.

Grasso, Marco (2007), "A Normative Framework of Justice in Climate Change", *Climatic Change*, Vol. 81, 223-246, <http://www.springerlink.com/content/a026112837118026/ > accessed 25 March 2007.

Grieco, Margaret and Hine, Julian (2008), "Stranded mobilities, human disasters: The interaction of mobility and social exclusion in crisis circumstances" in: Bergmann and Sager, 65-71.

IPCC (2007), *Summary for Policymakers*, in: *Climate Change 2007: Impacts, Adaptation and Vulnerability. Contribution of Working Group II to the Fourth Assessment Report of the Intergovernmental Panel on Climate Change*, M.L. Parry, O.F. Canziani, J.P. Palutikof, P.J. van der Linden and C.E. Hanson, (eds.), Cambridge: Cambridge University Press, 7-22.

IPCC (2007), Adger, W.N., S. Agrawala, M.M.Q. Mirza, C. Conde, K. O'Brien, J. Pulhin, R. Pulwarty, B. Smit and K. Takahashi, *Assessment of adaptation practices, options, constraints and capacity. Climate Change 2007: Impacts, Adaptation and Vulnerability. Contribution of Working Group II to the Fourth Assessment Report of the Intergovernmental Panel on Climate Change*, M.L. Parry, O.F. Canziani, J.P. Palutikof, P.J. van der Linden and C.E. Hanson (eds.), Cambridge: Cambridge University Press, 717-743.

Ingold, Tim (2000), *The Perception of the Environment: Essays in Livelihood, Dwelling and Skill*, London/New York: Routledge.

Kronlid, David (2003), *Ecofeminism and Environmental Ethics: An Analysis of Ecofeminist Ethical Theory*, (Acta Universitatis Upsaliensis, Uppsala Studies in Social Ethics 28), Uppsala.

— (2008a), "Mobility as capability", in: Cresswell and Priya Uteng, 15-34.

— (2008b), "What Modes of Moving Do to Me: Reflections about Technogenic Processes of Identification" in: Bergmann, Hoff and Sager, 125-154.

Leduc, Timothy B. (2007), "Sila dialogues on climate change: Inuit wisdom for a cross-cultural interdisciplinarity", *Climatic Change*, Vol. 85, 237-250.

—, "Climate Research, Interdisciplinarity and the Spirit of Multi-Scalar Thought", in this volume.

Lundgren, Lars J. (ed.) (2003), *Vägar till kunskap: Några aspekter på humanvetenskaplig och annan miljöforskning*, Eslöv: Symposion.

Lundgren, Lars J. and Sundqvist, Göran (2003), "Hur blir en förändring i naturen ett miljöproblem?" in: Lundgren, 27-72.

Mortimore, Michael J. and Adams, William M. (2001), "Farmer adaptation, change and 'crisis' in the Sahel", *Global Environmental Change*, Vol. 11, 49-57.

Northcott, Michael (2007), *A Moral Climate? The Ethics of Global Warming*, London: Darton, Longman and Todd.

Nussbaum, Martha (2000), *Women and Human Development: The Capabilities Approach,* Cambridge: Cambridge University Press.

Nussbaum, Martha and Sen, Amartya (eds.) (1993), *The Quality of Life*, Oxford: Oxford Univiversity Press.

Oelschlaeger, Max (ed.) (1995), *Postmodern Environmental Ethics*, New York: SUNY Press.

Ott, Konrad, "Ethical Claims about the Basic Foundations on Climate Change Policies", in this volume.

Paavola, Jouni (2005), "Seeking Justice: International Environmental Governance and Climate Change", *Globalizations*, Vol. 2, 309-322.

— (2008), "Science and Justice in the Governance of Adaptation to Climate Change", *Environmental Politics*, Vol. 17, 644-659.

Paavola, Jouni and Adger, W. Neil (2002), "Justice and Adaptation to Climate Change", Tyndall Centre for Climate Change Research, Paper 23, Oct., <www.tyndall.ac.uk> accessed 27 November 2006.

Page, Edvard. A. (2007), *Climate Change, Justice and Future Generations*, Cheltenham, UK, and Northampton, MA: Edward Elgar.

Qizilbash, Mozaffar (2005), "Dialogue: Sen on Freedom and Gender Justice", *Feminist Economics*, Vol. 11, 151-166.

Robeyns, Ingrid (2003), "Sen's Capability Approach and Gender Inequality: Selecting Relevant Capabilities", *Feminist Economics*, Vol. 9, 61-92.

— (2005), "The Capability Approach: a theoretical Survey", *Journal of Human Development,* Vol. 6, 93-114.

Sen, Amartya (1993), "Capability and Well-Being" in: Nussbaum and Sen, 30-55.

— (1999), *Development as Freedom*, New York: Knopf.

— (2005), "Human Rights and Capabilities", *Journal of Human Development,* Vol. 6, 151-166.

Stone, Christopher (1995), "Moral Pluralism and the Course of Environmental Ethics", in: Oelschlaeger (reprinted from *Environmental Ethics* 10, no. 2 (1988), 139-54.)

UNESCO (2009), *Learning and Knowing in Indigenous Societies Today*, edited by Bates, Peter, Chiba, M., Kube, S. and Nakashima, D., Paris: UNESCO.

Öhman Johan (ed.) (2008), *Values and Democracy in Education for Sustainable Development: Contribution from Swedish Research*, Stockholm: Liber.

ETHICAL FOUNDATIONS OF CLIMATE CHANGE POLICIES*

Konrad Ott

Program: This is a somewhat unusual paper. It is structured according to 48 steps that, seen in conjunction, should become an argument in favour of a specific concept in climate ethics, namely "Contraction and Convergence" (C&C). If the argument is adopted as being sound, it provides practical and political orientation in the field of international climate change policies. The overall argument enables critics to disagree in a more precise manner, rejecting certain steps. Such (and only such) rejections will make further debate more reasonable. The argument is intended to shed some light on the philosophical background and to an extent some of the political consequences. This argument presumes to be neither utopian nor to be like Minerva's owl that only flies in dim light after sunset (as Hegel said about philosophy). Rather it presumes to give support to a concept that has been regarded as being "utopian" a decade ago but has now entered the political arena. The case for C&C might turn out to be a paradigm case under the question of whether moral reasoning counts in the field of international relations. In what follows, I do not observe, at present such reasoning at the interface between ethics and politics. Rather, I argue as an ethicist hoping that this pattern of reasoning might be observable in international regime formation in the near future.

1. "The issue of climate change is ultimately an ethical issue." (Rayner et al.) Climate change is intrinsically related to moral issues. "Climate ethics" is a title for ethical reflections on those issues.
2. It has often been said that there should be a serious *debate* about moral and ethical aspects of climate change. If so, *discourse ethics* (Habermas) can serve as a procedural ethical framework that makes much room for substantial arguments. Being a variant of universalistic ethics, discourse ethics is appropriate with respect to global issues.
3. There is a "logical" superiority of *arguing* over *bargaining* (and, thus, of discourse ethics over rational choice theory). I am arguing, not bargaining about the follow-

* This overall argument is to be found at length in a report written for the German Federal Environment Protection Agency. Konrad Ott et al., *Reasoning Goals of Climate Protection: Specification of Article 2 UNFCCC,* Berlin: Umweltbundesamt 2004. Thanks to Christian Bartolomäus, Christian Baatz, Ulrich Hampicke, Barbara Muraca, Detlef Sprinz and Elisabeth Zelljadt.

ing claims. The claims are reasons which are to construe an overall argument in favour of a general conception that is known as "Contraction and Convergence" (C&C) (Meyer, see also the recent publications of Global Commons Institute).

4. Cost-benefit analyses (CBA) should be used as decision criterion only if there is nothing of moral importance at stake (Randall). If so, CBA should not be decisive in climate change issues. It should be used as one piece of information amongst many.
5. The idea of calculating an *economically optimal ("efficient") climate path* is misleading. There is no single utility function for mankind. Moreover, many value-laden problems are implicitly entailed in economic modelling (value of a statistical life, damage function, rate of discount, distribution of benefits and disadvantages, etc.). There is much highly contested ethics embedded in economics.
6. Economic approaches allow for discounting. Discounting implies the possibility of severe accounting errors in long-term environmental problems (see contributions in Hampicke and Ott). It remains highly doubtful whether global climate change falls within the scope in which "normal" discounting is to be applied. The debate about the Stern report is at its core a debate about the rate of discount.
7. Given number 5, one should discard the idea of an "optimal" solution and should instead rely on a *standard-price-approach* in climate economics. And if so, there is a division of labour between economics and ethics. Ethics has to substantiate standards by reasoning, while economics has to calculate how a standard can be met by minimizing opportunity costs.
8. Standards cannot be derived from purely scientific data, but must entail, broadly speaking, a "value-judgement" (IPCC, TAR). This is true for the ultimate objective of UNFCCC, as laid down in Article 2 of the framework convention.
9. The interpretation of Art. 2 UNFCCC should avoid naturalistic fallacies. Science as such can not derive political objectives. "Physical" criteria are always related to implicit judgements about the acceptability of long-term risks. This is true for a *Tolerable Windows Approach* or a *Soft Landing Approach* as well.
10. The term "dangerous" as is to be found in Art. 2 is intrinsically related to concepts of *risk* (danger, hazard) and uncertainty which are related to ethical concepts (gain, damage, precaution, distribution, risk aversion, care). There can be no risk theory without any implicit ethics (Skorupinski and Ott). If so, no interpretation of Art. 2 can avoid addressing ethical questions.
11. Value-judgements can be justified by argument. (It is misleading to shy away from moral reasoning just because such reasoning is not "scientific" but "ethical".)
12. Well-considered, reflective and balanced *judgments* are a third category between scientific derivations ("proofs") and arbitrariness. There are viable ethical strategies of *arbitrariness-reduction* in order to reach such reasonable judgment. One strategy is to look for *ethical convergence* (no. 17 ff.).

13. Therefore, a basic standard in climate policy might be substantiated on ethical grounds. Art 2 UNFCCC *addresses* such standard ("ultimate objective") but cannot *serve* as a standard without further interpretation and specification.
14. The *general structure* of Art. 2 is this: Three basic requirements (food production, natural adaptation of ecosystems, sustainable economic development) must be obeyed simultaneously in the overall courses of actions which are devoted to reach specific non-dangerous GHG-levels in a specific time frame. These three requirements are themselves open to interpretation.[1]
15. Art. 2 defines the ultimate objective of COP. The concept of a *political* objective entails the requirement that it should be specified in its decisive parameters (quantity, time-frame, actors, compliance, etc.). Specification could transform the ultimate objective into a true political objective.
16. Thus, a commonly shared ethical interpretation of Art. 2 FCCC (if there is any) could serve as a basic standard in global climate policy (Ott et al.).[2] It could serve as common moral ground. Such ground seems mandatory for successful negotiation in a Post-Kyoto framework that goes beyond pure strategic bargaining.
17. Currently, there is a plurality of ethical theories. This plurality is often seen as a fatal second-order-relativism. Nevertheless, it is possible to derive the same or similar judgments from competing ethical theories. If this logical possibility is actualized with regard to a certain moral problem, there is *ethical convergence*.
18. Ethical convergence generally counts as a strong rationale in (applied) ethics.
19. There is a broad consensus among ethicists that (members of) future generations have binding moral claims upon us. All so-called No-Obligation-Arguments are unsound (Ott). This holds true also for Parfit's so-called "future individual paradox" that rests on a ambiguity with respect to the concept of a "particular person". If this ambiguity is resolved, moral obligations hold with respect to future persons irrespectively of their (unspecific) individuality.
20. A basic distinction can be drawn between deontological and welfarist (utilitarian) ethics. In the case of climate change, either a) a given set of deontological principles must be applied to classes of more or less probable consequential impacts of climate change (*deontologic approach*); or b) the overall future hedonic changes which are involved in climate change impacts must be taken into account fully (welfarist approach).[3]
21. To deontologists, there is a fundamental *obligation to avoid injuries* in actions as well as in institutions. CO_2-emissions can be regarded as kinds of indirect and systematic injury to other persons (O'Neill). If the burden of climate change may fall onto vulnerable groups whose members did not contribute much to the problem, this should be perceived as a case of environmental *victimization*. Such victimization is morally repugnant. This holds true also with respect to climate refugees.
22. There is a strong *convergence* in utilitarianism and in welfarist approaches to-

wards low stabilization levels (Broome, Lumer). No utilitarian can agree to high stabilization levels since the overall and long-term hedonic changes are negative at a high confidence level.[4] The probable impacts of climate change are not compatible with the ethical idea to maximize the overall difference between good and bad states of the world in the course of time. If one adopts a variant of utilitarianism that gives more weight to units of pain than to units of pleasure, the obligation to avoid impacts of climate change that bring about pain and suffering becomes even more stringent.

23. Under a Rawlsian *veil of ignorance* (Rawls), which covers time and place of one's life-span, prudent persons would clearly opt for low stabilization levels. (This option is a rational "maximin strategy".) If one imagines a situation in which one's enemy determines one's place in a world of climate change one has good prudent reasons to wish strong mitigation.
24. *Neo-Aristotelian ethics* argues that a broad agreement of different groups of experts counts as a presumption in favour of practical reason. There is, in fact, a broad agreement about a ≈ 2 °C GMT target (compared to pre-industrial temperatures), a 0.2 °C/decade objective and a 450-500 ppmv CO_2-eq objective (see overview in Ott et al.).[5]
25. All approaches in *physiocentric environmental ethics* converge strongly towards low or very low stabilization levels because such levels can be derived directly from obligations toward natural beings. Climate change is seen as a paramount threat to global biodiversity. High emission scenarios are seen as morally repugnant.[6] Polar bears have become icons of climate change. As icons, they represent about 20 or even 30% of our planet's species whose risk of extinction might increase.
26. In cases of global and unique environmental problems, the choice of criteria of risk evaluation should be independent from one's individual degree of risk aversion. Most ethicists agree that in such unique cases with high damage potential a prudent society should better err on the side of caution. If so, more precautionary ("tutoristic") criteria should be chosen (Jonas). Tutoristic criteria speak in favor of a low stabilization level (Schröder et al.).
27. *Result of ethical analysis:* Most current ethical theories (utilitarianism, welfarism, Kantian deontology, Aristotelianism, Rawlsian approaches, Hans Jonas' ethics of responsibility, physiocentric environmental ethics) suggest low (or very low) stabilization levels. There is a strong and remarkable convergence in climate ethics.
28. This overall ethical picture would not change if it would be enlarged to the realms of Christian ethics. As paradigm cases one might read the lectures of the Archbishop of Canterbury or of the Ecumenical Patriarch. Recently, there are several declarations and studies from Christian churches which broaden this ethical convergence (Church of Sweden 2008, EKD 2009).
29. Those who wish to opt for higher stabilization levels *on ethical grounds*, have to

favour *either* contractarianism *or* neoclassical economic approaches of net present value maximization. If one makes such a choice, one is committed to these theories as a moral person.[7]

30. All things considered, 450 ppmv-eq CO_2 (or even less) seems to be a reasonable (political) number for an upper limit.[8] I wish to propose 450 ppmv CO_2 minus x as a reasonable judgement on a not too dangerous human interference with climate system.[9] Note, that 450 ppmv-eq CO_2 might not achieve a 2 °C target, if pre-industrial temperature are to be chosen as the benchmark and if climate sensitivity is about 3 °C, as the IPCC assumes in the Fourth Assessment Report. If one wishes to increase the probability that the 2 °C objective will be reached one should adopt the 450 number as CO_2 equivalents.
31. This objective (450 ppmv-eq CO_2) is almost out of reach. Very likely, we will face some overshot in the middle of the century. Nevertheless, it would be ethically wrong to conform the ultimate objective step by step to more "realistic" numbers which are to be expected under moderate or high emission scenarios. It is far better to know that humankind is goung to trespass the red line.
32. The requirement in Art. 2 to secure *food production* should not only focus on the dimension of global harvest but also on the dimensions of local or regional food security. The amount of global harvest does not imply food security. If so, we should carefully consider impacts of climate change for tropical and subtropical regions.
33. Freshwater is food. Freshwater supply will be a crucial field for adaptation.
34. The ongoing combination of a) destruction and fragmentation of natural habitats and b) ecosystem impacts of climate change which put an additional threat on such natural systems is regarded wrong by all current approaches in environmental ethics. Climate change puts a huge threat on biodiversity.
35. If so, the interpretation of the *ecosystem-adaptation*-constraint of Art. 2 should take into account the general objectives and values of the international conventions for protection and conservation.[10] This would strengthen the coherence of international environmental regimes.
36. There are different interpretations of the *sustainable-development*-constraint of Art. 2 due to different approaches in the sustainability spectrum (weak, intermediate, strong). There are sound reasons to adopt strong sustainability as a general approach (Ott and Döring).[11]
37. Thus, to "enable economic development to proceed in a sustainable manner" (Art. 2) means that economic development is to be restricted by a constant natural capital rule, a rule to invest in natural capital, and by well-known management rules for renewable and non-renewable resources as well as for natural sinks.
38. The wording of Art. 2 does not exclude the option of reducing dangers by means of *adaptation*. The ecological, cultural and social dimensions of adaptation should

not be underestimated by using simple examples as paradigm cases (as in building dikes or growing different crops). Mitigation will be a necessary condition of successful adaptation. Conservation of nature might be important as an adaptation-strategy (Ott 2005).

39. The relationship between adaptation and mitigation is neither "either-or" nor "the-more-the-less" but "the more mitigation efforts are undertaken the easier adaptation will become". This relationship should be mirrored in the criteria by which peoples are funded by an adaptation fund. Reducing emissions from deforestation, soil degradation, and the destruction of wetlands should be rewarded by adaptation funding.
40. It seems possible to judge different emission scenarios morally. The SRES-"story lines" can be made objects of ethically informed choice (WBGU).[12] There are clear advantages to Scenario B1 (Ott et al.).
41. If there is a low stabilization level, the total sum of future emissions in this century has to be limited. If so, remaining emission entitlements are to be distributed among parties. There are ethical and political questions about burden sharing and allocation schemes (Shue). There are different *equity-principles* to be applied to emission entitlements.
42. An argument for *egalitarian allocation principles* can be made as follows:
    - The atmosphere is a global sink for GHG.
    - This sink has to be regarded as being a global common pool good.
    - It is not adequate to zone the atmosphere (as zoning oceans) or to model it as distinct bubbles.
    - The atmosphere is to be perceived as "res communis", not as "res nullius".
    - Industrialized countries have used this sink in the past and, thereby, made the remaining sink capacity scarce. There is some historical responsibility.
    - There is a *presumption in favor of equality* in most theories of distributive justice. This presumption can be applied to the case of common pool goods.
    - Under this presumption the *burden of proof* falls on the side of unequal distribution of emission entitlements.
    - Unless this burden can be met by argument, egalitarian principles hold.
43. No. 30 in conjunction with no. 42 provides a rationale for *C*ontraction and *C*onvergence (C&C, Meyer).
44. If the overall argument given so far is considered to be sound, C&C has been justified on ethical grounds. This justification leaves room for adjustments and specification on the C&C-principle (population benchmark, transition period, emission trading).
45. If so, the following principles provide moral persons which *orientation* in climate ethics and politics:

- "ceiling" of 450-x ppmv CO_2 or (better) 450 ppmv-eq CO_2 eq (or even lower in the longer run),
- $x \leq 0.2$ °C/decade increase of GMT to allow for natural adaptation of ecosystems,
- scenario-line B1,
- strong sustainability,
- egalitarian (per-capita) emission entitlements as long-term objective.
- Rule of thumb: the more mitigation, the easier adaptation will be.

46. Therefore, political negotiations at the COP/MOP are to be perceived from a different perspective. Ethical orientation enables us to perceive barriers and obstacles (and drags) *as such*.
47. C&C is clearly incomplete. It must be enlarged as to cover the dimension of adaptation. There is room in C&C to recognize that there is a double obligation for rich nations to mitigate strongly *and* to give support for others to adapt.
48. There are several advantages of C&C over the Greenhouse Development Rights (GDR) approach. Under the criteria of environmental effectiveness, justice, acceptability, and feasibility, C&C deserves superiority over GDR (Kraus). Therefore, it seems both unnecessary and fatal, if the groups on the more progressive side of climate change policies were to split into two competing camps (C&C *versus* GDR).

Final outlook: Ethics is about solving practical problems ("How should we act?") by means of reason. It presumes to give orientation by making proposals. The argument presented in the previous 48 points proposes a specific "climate-ethical" concept as common moral ground for political negotiations. Without any common ground, climate policies will remain a "muddle through". Muddling-through looks highly pragmatic but it defers risks and nasty choices to the future. At the end of a muddling-through-period with increased global emissions, fatal shifts in the framing of the overall debate on climate change may incur. Facing failed muddling-through, geoengineering with sulfates or hard geostrategic prospects might look to be promising options. A new debate on geoengineering options has been launched recently (Victor et al.). Adopting and specifying the concept of C&C might be the best prevention against such means of last resort.

Notes

1 The more moral concerns (food security, species protection, "strong" sustainability) are embedded in the three basic constraints, the more stringent the obligation to reach low GHG-stabilization levels will be.

2 Without any such common ground, the COP probably will be trapped in "prisoners dilemma" situations as game theoretical modeling of climate policies has shown.

3 Hedonic changes can be specified by the WEHAB-approach (water, energy, health, agriculture, biodiversity).

4 Utilitarianism has always opposed discounting future utilities. Maximizing the net *present* value is not a principle of ethical utilitarianism.

5 To Aristotelianism, it seems acceptable to entail some assumptions about political feasibility in such objectives.

6 The fact that there has been natural climate changes in the history of Earth which caused extinction of many species does not entitle human beings to cause climate change. Such argument would be a naturalistic fallacy.

7 Contractarianism has severe moral shortcomings and flaws as general ethical theory. The "rationality" of the economic approach (maximizing one's expected utility under constraints) is opposed to morals.

8 Upper limits (ceilings) should not become targets. There is nothing wrong in maintaining headroom below such ceilings, if possible.

9 If one wishes to be sure to reach a "x = 2 °C" target GMT-increase one has to opt for lower numbers since 450 ppmv CO_2 (roughly 490 CO_2-eq) in conjunction with medium or high climate sensitivity is assumed to result in ≈ 2.5-3 °C increase in GMT.

10 Convention on Biological Diversity, Ramsar Convention, Convention on Migratory Species and others.

11 A general elasticity of substitution of no less than unity between man-made and natural capital is no longer an "educated guess" (Solow).

12 WBGU does not favor A1T, but uses A1T as an "existence proof" that a 450 ppmv CO_2-target is possible even under conditions of high economic growth.

References

Broome, John (1991), *Counting the Cost of Global Warming*, Cambridge: Cambridge University Press.

CAN (Climate Action Network) (2003), *A Viable Global Framework for Preventing Dangerous Climate Change*, COP-9, Milano.

Church of Sweden (2008), *Uppsala Interfaith Climate Manifesto*, Uppsala.

EKD (2009), *Umkehr zum Leben: Nachhaltige Entwicklung im Zeichen des Klimawandels*, Denkschrift des Rates der Evangelischen Kirche in Deutschland, Gütersloh.

Habermas, Jürgen (1983), *Moralbewußtsein und kommunikatives Handeln*, Frankfurt/M.: Suhrkamp.

Hampicke, Ulrich and Ott, Konrad (2003), "Reflections on Discounting", *International Journal of Sustainable Development*, Vol. 6, No. 1 (Special Issue).

Jonas, Hans (1979), *Das Prinzip Verantwortung*, Frankfurt/M.: Insel.

Kraus, Katrin (2009), *Contraction & Convergence and Greenhouse Development Rights: A critical comparison between two salient climate-ethical concepts*, Diplomarbeit an der Ernst-Moritz-Arndt-Universität Greifswald.

Lumer, Christoph (2000), *The Greenhouse: A Welfare Assessment and Some Morals*, Lanham: Rowman & Littlefields.

Meyer, Aubrey (1999), "The Kyoto Protocol and the Emergence of 'Contraction and Convergence' as a Framework for an International Political Solution to Greenhouse Gas Emissions Abatement", in: Hohmeyer, Olav and Rennings, Klaus (eds.), *Man-Made Climate Change*, Heidelberg, New York: Springer, 291-345.

O'Neill, Onora (1997), "Environmental Values, Anthropocentrism and Speciesism", in: *Environmental Values*, Vol. 6, No. 2, 127-142.

Ott, Konrad (2004), "Essential Components of Future Ethics", in: Döring, Ralf and Rühs, Michael (eds.), *Ökonomische Rationalität und praktische Vernunft: FS Hampicke*, Würzburg: Königshausen & Neumann, 83-108.

— (2005) "Climate change and the loss of biodiversity: an intertwined relationship and its ethical and political implications", *Report of the 13th Annual Conference of the European Environment and Sustainable Development Advisory Councils (EEAC)*, 32-37.

Ott, Konrad and Döring, Ralf (2008), *Theorie und Praxis starker Nachhaltigkeit*, 2. ed. Marburg: Metropolis.

Ott, Konrad, Klepper, Gernot, Lingner, Stephan, Schäfer, Achim, Scheffran, Jürgen and Sprinz, Detlef (2004), *Reasoning Goals of Climate Protection: Specification of Article 2 UNFCCC*, Berlin: Umweltbundesamt.

Randall, Alan (2002), "Benefit-Cost Considerations Should be Decisive When There is Nothing More Important at Stake", in: Bromley, Daniel and Paavola, Jouni (eds.), *Economics, Ethics, and Environmental Policy*, Oxford: Blackwell, 53-68.

Rawls, John (1971), *A Theory of Justice*, Harvard: University Press.

Rayner, Steve, Malone, Elizabeth and Thompson, Michael (1999), "Equity Issues and Integrated Assessment", in: Tóth, Ferenc (ed.), *Fair Weather?* London: Earthscan, 11-43.

Schröder, Meinhard, Claussen, Martin, Grunwald, Armin, Hense, Andreas, Klepper, Gernot, Lingner, Stephan, Ott, Konrad, Schmitt, Dieter and Sprinz, Detlef (2002), *Klimavorhersage und Klimavorsorge*, Berlin/Heidelberg: Springer.

Shue, Henry (1992), "The Unavoidability of Justice", in: Hurrell, A. and Kingsbury, B. (eds.), *The International Politics of the Environment,* Oxford, 373-397.

Skorupinski, Barbara and Ott, Konrad (2000), *Technikfolgenabschätzung und Ethik*, Zürich: Hochschulverlag.

Solow, Robert (1974), "Intergenerational equity and exhaustible resources", *Rev. Econ. Studies*, 29-45.

SRU (Der Rat von Sachverständigen für Umweltfragen) (2005), *Kontinuität in der Klimapolitik – Kyoto-Protokoll als Chance*, Berlin.

Victor, D.G., Morgan, M.G., Apt, J., Steinbruner, J. and Ricke, K. (2009), "The Geoengineering Option: A Last Resort Against Global Warming?" *Foreign Affairs*, March/April 2009, 64-76.

WBGU (Wissenschaftlicher Beirat Globale Umweltveränderungen) (2003), *Energiewende zur Nachhaltigkeit*, Heidelberg/Berlin: Springer.

WHAT ARE WE WAITING FOR?

CLIMATE CHANGE AND THE NARRATIVE OF APOCALYPSE

Stefan Skrimshire

INTRODUCTION

Nations and peoples are largely the stories they feed themselves. If they tell themselves stories that are lies, they will suffer the future consequences of those lies. If they tell themselves stories that face their own truths, they will free their histories for future flowerings. (Ben Okri)[1]

I perceive our ability to survive climate change as the enactment of an evolutionary narrative. (David Haley, ecological artist, Manchester Metropolitan University)

A Guardian article written in 2005 argued that, unlike reaction to fears about nuclear Armageddon twenty years ago, fears about climate change have not yet sparked a significant creative literary response. The lack of an "imaginative repertoire", argued Robert McFarlane, "reduces the ability of people to envision possible futures and consequently make better decisions today."[2] The diagnosis could be rephrased as saying that we have yet to find a way to creatively and usefully tell the story of climate change. But he also suggested that the *reason* for this dearth of imagination was that "climate change is not – not yet – apocalyptic in its consequences. Apocalypse comes swiftly and charismatically... [b]y contrast, climate change occurs discreetly and incrementally, and as such, it presents the literary imagination with a series of difficulties: how to dramatise aggregating detail, how to plot slow change."[3]

How should we view this judgment today, almost four years on? Uses of the word "apocalypse" in reference to climatic tipping points are increasingly common and normally accompanied with images of natural and social disaster. The four horsemen of conquest, war, famine and death appear no longer to be the preserve of millennialist scaremongers. Mcfarlane's point, presumably, is that even today climate Armageddon is still not the same as nuclear Armageddon. The complexities and scientific uncertainties of the former do not allow it the imaginative purchase that was achieved by someone having the obliteration of human life only the pull of a trigger away. Nevertheless I wish in the course of this chapter to raise a question mark over his two assumptions: 1) that climate change is slow and incremental; and 2) that apocalypse comes swiftly and charismatically. For whilst there may be *some* truth to these interpretations, both rely on a simplistic and potentially misleading understanding of the experience of crisis through time. To the first point, therefore, I wish to respond in this chapter that whilst it may be true that climate change does not generate the same sort of fear as the nuclear threat of Mutually Assured Destruction, this distinction may simply reflect the perspective of the privileged. Those in the affluent Northern Hemisphere still able to cope with mild changes in the weather are unlikely to rave about millennial end-times until the weather itself becomes apocalyptic, as exactly happened with religious commentators on Hurricane Katrina in 2005. We may live, as Frederick Buell has put it, in a culture of "hyperexuberance" that not only endures climate change but flourishes in it.[4] But this is for reasons of the cultural normalisation of a certain *type* of crisis, not the nature of climate change itself or how serious to our survival it is. To the second point, I wish to make it clear that if we are to categorise environmental catastrophe as "apocalyptic", it is high time we refined our understandings of where that notion comes from, and not only for reasons of etymological pedantry. For a brief examination of the functions, effects, and beliefs associated with apocalypse as a mode of revealing something divine previously hidden to earthly beings, also suggests some missed opportunities in telling the story of climate change. In short, I wish to argue that we wait at our peril for a "swift and charismatic" climate apocalypse to be able to define a definitive narrative of climate catastrophe by which human actors might position themselves. Typically, this is generated today by defining a point at which the crisis is so severe that humans have failed; are doomed; or, in another terminology, have eschewed (environmental) salvation. Instead, by taking heed of some *dramatic* aspects of the telling of apocalypse stories, we learn that its essential feature is the radical openness of the future, and its invitation to see in the days to come an imperative for resistance even beyond the tipping point of climate salvation.

The task outlined above will explicitly attempt to bridge a gap too little attended to between theological interpretations of apocalyptic in the Judeo Christian tradition and contemporary cultural appropriations of that word. Is there some connection that understandings of the former can teach the latter? I should be clear, however, that my intention is not to draw parallels in the sphere of religious belief between audiences of apocalyptic texts in the first century Palestine and those interpreting climate change (such as Christian millennialist responses to Katrina, for instance) in the twenty first century, though such work is valuable and needed. This study is concerned only with the connections that might be made between the original function of apocalypse (if indeed we can make such a claim) and its secular incarnation in the imaginary of climate disaster. For undoubtedly there is something retained in the cultural imagination that relates to religious perceptions of impending crisis. But precisely*what* our relationship is with this crisis is in question. How is it lived, understood and responded to? Or, to put it another way, what are we waiting for? The paradox facing activists and campaigners confronting climate change lies in how we interpret multiple crises whose effects are *both* incremental, gradual and imperceptible to the majority (such as melting of glaciers over decades), *and* also apocalyptic in the populist sense: both the visible appearance of natural disasters and social breakdown, and the passing of tipping points of global warming. More than a metaphor for how *bad* the outlook is, then, apocalypse tells us perhaps something much more revealing of our predicament and how we might respond to it. For at the root of these visions of end-times is this blurring of the boundaries of that which must come to pass, and that which must be resisted. More than a simplistic dualist perspective in which the righteous are saved and the unrighteous are punished, apocalypse can be seen as an essentially dramatic rendering of the processes of an originally good creation: in which corruption is purged and balance is restored to a world that has become unstable and unhinged.

In the context of climate, ecology, and the life of the earth, we have been indebted for centuries to the ability to tell stories of birth, death, and renewal. From oral to textual traditions, moreover, it is frequently the ritual enactment of such stories that give them life-giving and sense-making qualities. The active, "making" element derived from the concept of mythic truth or "mythopoesis"[5] is present in creation myths far beyond the western traditions, but is also an element of the Jewish contribution of the apocalypse myth. It certainly suggests a way of telling the climate story with more depth than Mcfarlane's call for instilling fear into its audience. This is not just because fear tends to paralyse rather than engage, though this is certainly true. It is because fear is not an experience that captures the complexity of even the severest accounts of dangerous climate change. It is easy to point out to Mcfarlane that climate change was anything but slow or incremental for the residents of New Orleans who stood in the path of Katrina exactly one month before his article was written. But this highlighting of a singular event – for those in New Orleans – raises precisely the problem. The cultural understanding of climate change is diffuse. It doesn't threaten everybody simultaneously in time and equally across space, as may have been the case during nuclear standoffs of Mutually Assured Destruction. We are reminded time and again, that the poor will suffer first and hardest – a point that Michael Northcott illustrates convincingly in this volume –, and we begin to become wary of exactly which members of the human race the proposed techno-fixes will save. We might say that climate change is still a story told internally, within boundaries, classes and traditions, but not outwardly, uniting values, ideas, and humanity itself. Our approach to story telling will therefore demand a new approach to crisis rhetoric itself. We must make sense of the end that has come already for some (the evacuees of Tuvalu, for example); is feared and anticipated for others, and is denied by the rest. These accounts of the end must somehow cohere and give us meaning. But even more: they must be able to give us bases and foundations for meaningful action. How are our disparate experiences of the feared end shared across experiential boundaries, across time and cultures? What leads some to retreat to a state of isolation, others to build movements of resilience in communities, others to hasten the end in the style of millennialists and chiliasts?

Narrating apocalypse: the ambivalence of endings

We have wrongly expected the end of the world would provide the high drama we believe commensurate with our raging passions, our bold aspirations, and our central importance to the universe – we are worthy of a big bang, not merely a whimper. And let me be blunt: by holding out for that noisy demise, we can pretend we haven't been expiring by inches for decades. (Andrew McMurray)[6]

The language of apocalypse in the media appears to capture something quintessential about climate reporting that has obviously surfaced since Mcfarlane wrote his article. Some newspaper headlines serve as an example: "Countdown to the Apocalypse: Scientists Change Doomsday Clock" (Fox News);[7] "Apocalypse Now: How Mankind is Sleepwalking to the End of the Earth" (The Independent). If we are searching for that element of apocalyptic discourse that might allow us more creatively to tell the story of climate change, it is necessary first to understand in what sense we might understand it as story, or narrative, in the first place. According to Paul Ricoeur, narrative implies symbolic invention. Its essence, like other semantic devices, is innovation: bringing new meaning to experience otherwise, we might say, hidden from view in ordinary perception or discourse. With the insertion of plot, "goals, causes, and chance are brought together within the temporal unity of a whole and complete action."[8] We tell stories in order to allow the future to make sense of our present. This can be appreciated as much for the happy ending of the fairy tale as for the sense of loss experienced in tragedy: in either case the experience of the present is given fuller meaning by reference to a future. But the reason that this principle is so pertinent to apocalypse is its emphasis on the uncovering of the new. As revelation or disclosure, *apocalypsis* means exactly this: "the new thing – the as yet unsaid, the unwritten".[9] It reveals by mixing temporalities – the untimely incursion of the future into the experience of the present; the bringing of a crisis from heaven to earth. Out of the collection of apocalypses drawn from intertestamentary (3rd Century BCE-2nd Century CE) Palestine and the Near East, this aspect remains fundamental to the genre: apocalypses are visions granted to privileged human recipients of heavenly truths.[10] In the book of Daniel those truths appear in the form of a series of dreams; in 1 Enoch as a series of Dream Visions, and in the book of Revelation as visions granted to John through an angel emissary. Apocalypses, in both an etymological sense and in its new cultural appropriation, are thus experiments in disclosure. Where they depart from the prophetic tradition in Judaism is in heightening the crisis wrought by the corruption of the present. Both spatially and temporally, the present world is, in texts such as Revelation, so far removed from the world that redeeming it with the heavenly vision of the world as it *should be* is to completely (and violently) overturn the reality of the present world.

The apocalypse texts developed this theme of the disclosure of truth precisely through the very act of narrating: of inserting plot imaginatively into the minds of its audience. It has been argued, for instance, that the book of Revelation dramatically enacts a theodicy of sorts by its enactment of a series of punishments on the unjust of the earth. Through a period of tribulation the forces of light overcome those of darkness and the efforts and suffering of those who held to the cause of light are vindicated.[11] In this process the disclosure of truth is twofold. The revelation to John itself represents the privileged access of human beings (however cryptically encoded) to ultimate or superior truth, or *aletheia*. And the characters that are vindicated in the narrative are those who cling to truth, who resist the mesmerising allure of the "beast".[12] Its apparent function through both aspects would appear to be not only a dualistic separation of the cosmos into good and evil but more essentially a call to discernment of truth from falsity, an opening of one's senses, and in particular the power of *seeing*, of this revelation being in some sense an awakening from blindness into sight. Michael Northcott's chapter proposes precisely one aspect of this healing visionary aspect of apocalypse, through the symbolism of *trees* in the Christian tradition (and in the book of Revelation itself) as truth-tellers. That is, both the truth of our own self-imposed destruction, and the hope in continued life.

It is with this emphasis on clear vision that we should approach the most obvious parallel sought after in apocalypse for interpreting climate change. What, we should want to know, is it to teach us about the *end* of this particular planetary or cosmic story? Revelation's culmination in the temporal details of the end times – in particular the inauguration of the millennium as penultimate end-period for humanity – leads most interpreters to look to apocalypse as the sign of impending finality. The end is specifically *near*: not just at some point in the future, but near, close at hand, and its signs should be interpreted and acted upon with a sense of urgency. But what we learn from apocalypse more generally is that the importance of ending need not have a clear sense of temporal finality: the apocalypse genre shows no consensus on eschatological intent. Where reference is made either to the resurrection of the dead (Daniel and Revelation) or the inauguration of a new age, "future expectation" has been taken by some scholars to mean the transformation of the present hopes of the earthly community. The dominant feature of Jewish apocalypse is thus a *transcendence* of death to a "heavenly form of life" as opposed to its reduction to a temporal sequence of events.[13] Such a view does not go undisputed, and certainly the most vivid accounts of future catastrophe visited upon the world that survive into contemporary culture are the very features of Revelation that speak of the physical and planetary transformations that can be expected and envisaged: the burning of the earth; sea turning to blood; locusts torturing humans; the decimation of human populations.[14] Are these signs transformations to be feared and anticipated or identified with the corruption of the present? Here indeed lies the agonistic ambiguity to the apocalypse genre as narrative, as both tragic and comic unfolding of the human drama: the promise of destruction for the wicked and salvation for the righteous. But the ambiguity is a temporal one, too. For apocalypse discloses a future that is both to come, and which transforms the present through its revelation of divine truth. Through dramatic envisioning of the events that are to come, the seer, John, enacts the very reality of those events one is to expect: "I saw a new heaven and a new earth, for the first heaven and the first earth had passed away, and the sea was no more".[15] Such a sentiment presages the more developed messianic eschatology in the Christian tradition. The message to the first Christian community was one of the kingdom having been established, of the old powers having been crushed, and yet also of the kingdom awaited, yet to come. In St Paul's eschatology, it was charged with interpreting this confusing mixing of temporalities as the meaning of the fullness of time, the nature of its "fulfilment" or "supersession".[16] Apocalypse represents this contraction of time, its reconfiguration according to a new reality, the *incursion* of the future into the present. But this provides an interesting angle on the diverse ways in which temporality is played with in our own discourses of the end, whether through film, media rhetoric, or common speech. For to envisage the end in graphic and visionary terms is not simply to predict calamity but to taste it *now*. It means to recognise its very anticipation as transformative of the present.

Just as there is ambiguity as to the eschatological content of apocalypse – whether and how it refers to a divinely ordained future – there is also ambiguity over whether it refers to end as terminus or simply to its irrevocable degeneration. Within the apocalypse genre this is a feature of the transformations in literary appropriation since the latter half of the twentieth century. Apocalyptic novels from this period normally envisage a radically altered state of creation or civilisation *after* the event of calamity.[17] This gives rise to the phenomenon of post-apocalypse, from works like Walter M. Miller's *Canticle of Leibowitz*, or more recently Cormac McCarthy's *The Road*. But more generally we can see that modernity's originally apocalyptic love affair with teleological progress – what is *becoming* and where we are all headed – has been subtly perverted through the obsession with what *continues* in perpetuity. What this means is that our recently rediscovered love of apocalyptic metaphors through climate reporting occurs against a backdrop of this sense of the permanence of crisis. Jean Baudrillard is the traditional spokesman of this revelation. Throughout the eighties and nineties he reminded us just how far we have digressed from the triumphalist idea of progress' crowning achievement in the end of history. In its place we have, through a culture of consumption, a society possessed by its experience of the present, devoid of future projection. The meaning of apocalypse changes in this postmodern context to a crisis of the never ending: "Our Apocalypse is not real, it is *virtual*. And it is not in the future, it is *here and now*."[18] Baudrillard's critique of the "hyperreality" of meaning in mass media society is important for our understanding of climate reporting, since it suggests that what is revealed to us about impending crisis is something both out of reach (untouchable, never arriving) and bound to the realm of aesthetics as opposed to ethics. We all know about the seriousness of climate change – its reality is in *some* sense ubiquitous – and yet feel powerless to do anything about it. Mediatised apocalypse is, as Andrew McMurray has put it, "Armageddon by surplus meaning, a drowning in honey".[19]

What could possibly be revived from this cynical backdrop to help us understand our climatic apocalypses? We should want to affirm, for one, that despite a general cultural block on imagining the future, apocalypse has a life of its own. This is certainly true from the historic inability for failed predictions of the end of the world to dent its credibility. As Ricoeur has said, this shows a qualitative transformation of the apocalyptic model. For in place of the insistence of temporal prediction – that some *thing* will take place by a certain *time* – multiple failures of such a literalist approach (with famous examples in the history of Seventh Day Adventism and the "Great Disappointment" in 19th century America) generated a switch to its interpretation as a sort of state of permanent expectation of catastrophe:

> from imminent to immanent. The Apocalypse, therefore, shifts its imagery from the last days, the days of terror, of decadence, of renovation, to become a myth of crisis.[20]

Applying this to climate reporting we notice a similar transformation, and may need to decide how to best direct it. For the problem with reporting climate apocalypse in predictive (or, more precisely, *projective*) terms of the last days is not with its appeal to the seriousness of the reality before us. The problem, rather, will be with how to maintain its credibility *beyond* the prescription of certain definitive tipping points, as a threat yet to be averted. The problem is demonstrated well in a recent campaign called "100 months": entering the website reveals a live clock audibly counting down the seconds. Above the clock are the words: "we have 100 months to save the planet. When the clock stops ticking we could be beyond the earth's tipping point, the point of no return."[21] Notwithstanding its use of the "could be" clause, the rhetoric appears to suggest that beyond a given, singular point – a concept at odds with many scientific understandings of tipping points, as I shall argue – redemption is out of our hands. What then? What narrative means do we have to extend the ethical import of global crisis beyond the tipping point? What resistance can there be when our sense of loss is irretrievable, past a point of no return?

The alternative is to understand that, as with much great tragedy, "crisis replaces the end, where crisis becomes a symptom of the invalidation of the paradigm itself."[22]

The function of apocalypse in *this* sense is to paint crisis not as a period of passive judgment but as a period of disclosure to ordinary people of "things hidden since the foundation of the world".[23] Apocalypse, then, might be viewed more generally as the emergence of secrets, the overturning of old ways of seeing and perceiving the world. Through the drama of apocalypse the world is really changed, turned upside down, and this occurs through being given a new sight, traditionally through the intermediary of an angel, as in the case of the New Testament apocalypse: "he made it known by sending his angel to his servant John".[24] It opposes a linear reading of history: to see apocalyptically is to affirm the strangeness and *otherness* of the world. The point has not been lost on philosophies of discourse itself: In the philosophy of deconstruction we find the principle of rupture appear as this same celebration of strangeness, the same rejection of synchronicity and chronos time. Thus Jacques Derrida's counterpart to St Paul's "the time has grown short" can be seen in the use of Hamlet's expression, "the time is out of joint, *hors de ses jonds*" (off its hinges). The aim is to show "the impossibility of thinking or rather of conceiving of the contemporary, of synchrony".[25] What is the act of announcement, he asks:

> What effect do these noble, gentile [*gentils*] prophets or eloquent visionaries want to produce? With a view to what immediate or postponed [*ajourné*] benefit? What do they do, what do we do in saying this? For whom do we seduce or subjugate, intimidate or cause to enjoy, to come?[26]

The questions could be placed in all seriousness to ourselves, the self-appointed messengers of a new global crisis. But for Derrida there is something primal to western thought itself in John's act as the messenger, this role of being the favoured dispatcher of revelation and denouncing the false ones, the "impostor apostles".[27] And Derrida's concern is of course to show how a deconstructive reading affirms the possibility of visionary predictions of a completely different global – or ecological – order. To announce "the new" is to flood the production of cultural imagination, or in other words to break out of its ideological boundaries and imagine new possibilities, ruptures, and breaks with the past. It takes full advantage of the post-modern fragmentation and diversifications of discourses and truths against the homogenous voice of authority and tradition:

> by its very tone, the mixing of voices, genres, and the breakdown [*le détraquement*] of destinations, apocalyptic discourse can also dismantle the dominant contract or concordat. It is a challenge to the established admissibility of messages and to the enforcement or the maintenance of order [*la police*] of the destination, in short to the postal regulations [*la police postal*] or to the monopoly of the posts.[28]

We need not disappear too far into deconstruction to appreciate its major recommendation: that we crack open the dramatic story telling power of apocalypse. The language of climate catastrophe can also fall for the hegemony of one vision, one sense, one meaning: we are on this path, and when it reaches a certain point, we are all finished, end of story. But as I have already asserted, climate crisis means a great many different things to a great many people, as does the sense of hope in the future – what kind of future do we want to preserve? And in any case, it is, perhaps, for some people, *already* finished: those for whom, like the residents of the disappearing small islands of the world, a point of no return has already been reached. To assume a *singular* point of failure in the fight against catastrophic climate change betrays a fundamental impulse to find reasons to resist in the most despairing of situations for goals that will differ for different people. As such it falls into the very trap that deconstruction warns against: the division of the world into binary opposites. It has long been thought that the tendency to divide the world into black/white, good/bad, rational/irrational is virtually universal to human cultures.[29] Moreover it is this division that has defined our ability to separate, control, and have power over. It also defines our division of time, history, and the sense of inevitability we inherit, seeing history as either won or lost; epochs as passing from one to the other. This rejection of binary opposites actually opens up a way to communicate crisis more creatively and effectively. In apocalypse we have the future in the present, but in more than a figurative sense. In ancient biblical apocalypse the old law is not just brought back to complete the story, reaffirming genesis as origins through apocalypse as destiny, but in order to break with the old order and inaugurate a "new heaven and a new earth". Non-canonical apocalypses such as the "animal apocalypse" of 1 Enoch (written between 3rd and 2nd centuries BCE) have made this concept of superseding old divisions and laws of separation (such as food laws) explicit through its visions of the new breaking of taboos, the mixing of originally distinct elements.[30] My own interest, however, is the way in which apocalypse presents the opportunity for seeing the certainties of the present as disrupted by the incursions of the future. In this sense the opposition of binaries matches with Derrida's opposition of "synchronicity" and the strict view of historic epochs, of a historic fatalism and certain knowledge of the way the future will lead us.

Apocalypse purposefully confounds the *temporal* orientation towards the future that its listeners will be accustomed to. The gripping reality by which the visions occur to John of Patmos are not simplistic predictions of the future. And revelation is fundamentally not passive communication: it draws the hearers into identifying their own part in the secret unfolding destiny of the universe. The secret to the text is in its dramatic performance.[31] Aristotle's theory of tragedy explained the process of the cathartic journey of its listeners through the narrative exploration of their despair in order to allow them expression. Whilst Revelation also contained tragic elements and so arguably a cathartic rendering of the punishment for Babylon (Roman Empire), it also contained comic elements in the resolution of crisis for the righteous by the end.

Those who are faithful to "the book" can identify with the saved of that dramatic narrative. Its violence, in other words, is not arbitrary, but engaging: it offers its spectators a "role in the cosmic drama".[32]

Clearly, apocalyptic texts are not to be sought for analogies to our reading of the truthfulness of climate catastrophe. The tragic unfolding of climate related disasters affects the lives and wellbeing of many millions of people in the present, and is only compared to the religious myths of ecological destruction (starting with the flood and ending with the apocalypse) in symbolic terms. But the understanding of crises yet to come – how they constitute us, push us into new forms of living and believing in the future – depends entirely on how we communicate it to ourselves and others, and therefore really cannot avoid a discussion about our most ancient of crisis narratives. Apocalypse constitutes the western psyche not only for its religious longevity in millennial movements, but for its power to draw listeners into a dramatic engagement with the reality of total upheaval. As Oliver Smith has put it, such an emphasis requires the tradition of *pathos* as opposed to *ethos* in the emotive engagement with such a text – one that fully involves the whole person of the listener.[33] The telling of the meaning of the future – its revelation – also means enacting the vision we wish to be part of. Prediction of radical change becomes synonymous, or at least conceptually entangled, with the enactment of change. As O'Leary puts it: "apocalyptic succeeds or fails with its audiences to the degree that it persuades them of their situation within the particular historical pattern of temporal fulfilment represented in its mythic imagery."[34]

The Climate Abyss: Can We Look Beyond the Tipping Point?

A friend and fellow activist said, "What will it take for you to finally call it an apocalypse? The death of the salmon? Global warming? The ozone hole? The reduction of krill populations off Antarctica by 90 percent, the turning of the sea off San Diego into a dead zone, the same for Antarctica? How about the end of the great coral reefs? The extirpation of two hundred species per day? Four hundred? Six hundred? Give me a specific threshold, Derrick, a specific point at which you'll finally use that word." (Derek Jensen)[35]

In what ways does climate change discourse share some of the paradoxes and questions of ancient apocalypse, and can it learn anything from that relation? An obvious point of interest is the way in which violent and shocking images are used to shake people out of some presumed complacency with regard to climate change, just as the heightened portrayal of a world divided by the unrighteous and the righteous might have bee thought to do for Christian and Jewish populations battling Roman imperialism. There are many reports now emerging that contest the efficacy of such a strategy today, citing the fact that people more often find the magnitude of the problem too big an obstacle to viewing themselves as any sort of agents of social or ecological change.[36] But it is to a more complex point of comparison that I want to turn, and that is very notion of temporal ambiguity I have discussed above and the way in which it relates to climatic "tipping points" or "tipping elements". British climatologist Timothy Lenton defines tipping elements as those "subsystems of the Earth system that are at least sub continental in scale and can be switched – under certain circumstances – into a qualitatively different state by small perturbations."[37] A number of these can produce predictions of points at which attempts to stem global warming will become impossible: numerous positive feedback loops will produce unstoppable momentum in the earth system – a point of no return for the escalation of global warming, drought, and rising sea levels.

It is to this scenario of irreversible global warming that the rhetoric of human extinction normally refers. The term has predictably become a key reference to the politically sensitive notions of transforming mitigation to adaptation in policy language: if it is too late to halt runaway global warming, by what reasons might we endorse cutting carbon emissions – might it be simply too late to make meaningful change? What is the rhetorical function of the language of extinction? Is it a strictly philosophical interest resurrected from environmental ethics that we are to consider the import of the destruction of our own species? Are we to imagine that the *total* elimination of human life is a scenario so morally repugnant it will force massive cultural change, and if so does this leave a scenario of vastly diminished human activity a bearable evil? Consider, for example, the angry reaction of campaigners to the words of a chief scientific advisor to the UK government, who recommended that it prepare for 4 degrees extra warming above pre-industrial levels, despite its own commitment not to exceed 2 degrees. Such an admission amounted to, in the words of one commentator "in the immortal words that Chief Seattle probably never spoke, 'the end of living and the beginning of survival' for humankind. Or perhaps the beginning of our extinction."[38] "It's basically the end of the world", adds another article commentary on the same scenario.[39] In other words, government ministers themselves are describing a scenario of *process* rather than event (and recall the lack of such a distinction by Mcfarlane's article at the beginning) that may be equated with finality – the point at which all is lost. Some climatologists have already put dates on such inevitability, with one famously stating in 2005 that humanity faced extinction before the end of the century.[40] But the ethical significance of a prospect of an end to the human species is not as black and white as ethicists would like to be able to discuss. In relation to climatic tipping points in particular, we are aware that the sense of finality approaching human life is much on a par with the never ending apocalypses of postmodern fiction. It is not the end of human life we fear, but its irretrievable transformation into something we would not associate "meaningful life" with. In 2006 the scientist James Lovelock stated that we have already passed the crucial tipping points in the earth system that decree unstoppable warming: "Before this century is over, billions of us will die, and the few breeding pairs of people that survive will be in the Arctic where the climate remains tolerable."[41] In Mark Lynas' hugely popular thought experiment *Six Degrees*, he also concludes that even in the most extreme of warming scenarios, a wipe-out of *all* humans is unlikely:

> Unlike the Permian terrestrial animals, we can stockpile enough food in preserved form to last for many years. We can create artificial atmospheres to isolate ourselves from what is going on outside. We could one day even set up colonies on other planets. And yet...[42]

Whilst not portraying a literal end to the human, such scenarios do in any case conjure a situation that many would want to associate as an *end* in perhaps an originally apocalyptic sense: a world so transformed it bears little relation to the one we are so concerned to preserve in the present.

Such a portrayal of climate change introduces a complex environment of approaching and passing timescales for what, depending on one's view, is (for example) achievable, utopian, impossible, worth striving for, etc. in the context of mitigation. A host of culturally sensitive questions also impose themselves in this regard: to whom is the notion of a "no return" relevant? To the human species as a whole, or to a notion of "civilisation" we deem worthy of survival? And what of the value of non-human life beyond the passing of the human – does this feature in our calculations of the passing of meaningful tipping elements?[43] But underlying this cluster of questions proper to moral philosophy is a concern, not unlike the discussions surrounding the function of apocalypse, with the perception of crisis as it is both *anticipated* and *suffered* in the present. In either case, we want to know: how does a narrative of crisis transform our current ways of thinking and acting in the world, and how dependent are we on a voice (of authority, credibility, etc.) that tells us which crisis events are certain to pass, and which are unavoidable? An immediate response might claim that the answer is simple: avoidable crises give us reason to strive like never before; whilst unavoidable crises give us new imperatives to adapt to new and difficult situations. But on closer inspection the responses would seem not to conform as easily as this. Consider, for example, the premise behind *The Age of Stupid*, a documentary-style film released in the UK in January 2009. The film has one subtly unique methodology that marks it out from other climate change documentaries: it narrates the story of global warming from a state of retrospective failure. An archivist from the future, played by the actor Pete Postlethwaite, looks at footage from the past (our present) that signalled the urgency of global warming. He adds, with a note of incredulity: "Amazing. What state of mind were we in, to face extinction, and simply shrug it off?"[44] From interviews with the directors, we learn that the narration of failure is a rhetorical shock tactic – a means of shaking viewers out of collective apathy. But is it *only* this? Or is there a more controversial cathartic emphasis here on the act of mourning? Perhaps ambivalence amongst activists to the uncertainty of what exactly is lost reflects the propositional and ambivalent nature of hope itself. Indeed it may be possible to express both a suspicion that in some sense we are doomed, or our best efforts to contain runaway climate change come too late, and also the feeling that we can stop "it", whatever that task ahead of us is. As Arran Stibbe notes, populist climate science reporting increasingly reports from within this troubled zone of both citing evidence for how curbing adequate carbon emissions is still possible, and being fully aware that what this demands of society is so extreme as to be virtually inconceivable. The rhetoric, he argues, increasingly resembles that of cancer patients "who are so obsessed with the fight that they cannot plan for the possibility of anything other than recovery."[45]

Environmental groups have long predicted the eventual collapse of ecosystems and its social consequences: groups such as Earth First! are often labelled as secular apocalypticists for, in the early days, putting a date on the collapse of ecologies supporting human civilisation.[46] But the novelty represented by *The Age of Stupid* actually draws upon the more nuanced understanding of apocalypse that I have outlined above: its ambivalent status as both here and not here, now and not now. This may in fact be a feature of our everyday ability to maintain true contradictions as simultaneous beliefs: we both hope for what is possible to achieve, and mourn its failure to arrive. Similarly, we must look for ways in which climate change discourse communicates simultaneously that a threshold has already been reached (some amount of warming is already unstoppable) *and* that a crucial threshold (a point of no return) is still to be resisted. There are straightforward scientific reasons for this: some tipping points are believed to have passed, others not. But the ethical import of the mode of communication is vastly underestimated. It is the rhetoric that steers a course between "too late", "almost too late" and "too late within this political time frame for action" that will dictate the ability for people to do anything about climate change. The language of tipping points raises the complex interconnection between these temporal states of anticipation because of its own inherent sphere of uncertainty. It is known, for instance, that up to now humans have contributed around only 0.8 degree planetary warming rise. But it is also estimated that there is a time lag of around 40 to 50 years between carbon emissions and global warming, leading some to conclude that the warming to which we are committed already pushes us over the point of no return in global warming.[47] But given the uncertainty over whether or not such a singular point has arrived, what is the most appropriate ethical response? To assume that our life on the planet is going to be severely curtailed, or to act as if it is not? If there is uncertainty, shouldn't we always act *as if* the time had not run out, in order to maximise our chances of survival? As NASA scientist Gavin Schmitt puts it, "Much of the discussion about tipping points, like the discussion about "dangerous interference" with climate often implicitly assumes that there is just "a" point at which things tip and become "dangerous". This can lead to two seemingly opposite, and erroneous, conclusions – that nothing will happen until we reach the "point" and conversely, that once we've reached it, there will be nothing that can be done about it. i.e. it promotes both a cavalier and fatalistic outlook."[48] This, however, lies at odds with the sense of moral urgency required by many environmental activists. Plane Stupid activist Leo Murray writes, for example, that despite the inherent uncertainty in latent warming effect,

> These are extraordinary times. Preventing runaway global warming is the single most important task in all of human history – and it has fallen to us to do it. If we don't, then everything else we work to achieve in our lives will be destroyed, or become meaningless. Those who came before us didn't know about this problem, and those who come after will be powerless to do anything about it. But for us, there's still time![49]

But this huge responsibility appears to ignore another tacit agreement amongst environmentalists, which is that the world is already in some sense irrevocably changed. A sense of mourning and loss is therefore not misplaced: the World Health Organisation estimates that over 160,000 people already die each year because of climate change. We are also currently in such a rapid state of species extinctions that some call it one of only six "mass extinction" events known to history.[50] The same principle goes with the common association of final human loss not with literal extinction, but with the wipe-out of human civilisation: Hear Leo Murray again:

> 'Humanity' may survive this. But what will 'humanity' mean in a world where countries which remain habitable – like Britain – use most of our remaining resources fighting to keep out the starving millions who can no longer live in their own countries because of what we have done?[51]

A good question indeed, but one which should also acknowledge what humanity and civilisations means *today,* in the light of the irreversible losses already incurred by anthropogenic climate change.

How might the dilemmas above find any clarification in our previous analysis of apocalypse? The point of intersection appears to be the point within climate change discourse in which agency – the possibility of change, the ability to act – becomes swallowed up in some larger meta-narrative beyond people's control. Arguably, we are witnessing such a temptation through the popularity of what Lynas calls "geological fatalism". There is a perceived shift in public discourse from environmental action to an acceptance of the inevitable death of humanity, earth, and life itself, within a wider cosmological picture. It is arguably present in the recent explosion of thought experiments on the survival of life on earth *after* humans have (for whatever reason) disappeared: thought experiments such as Alan Weisman's *The World Without Us*; the National Geographic's *Aftermath: Population Zero* and the History Channel's *Life*

After People. Each documents a desire to see *beyond* the point of no return, to know what continues in spite of our absence. What might be the social function of such a perspective? One view is that it may reorient the moral priority afforded the lifespan of humanity in the interests of a wider intrinsic value of the biosphere, and beyond, from which our life is generated. As Heather Eaton has put it, lessons from truly understanding the scope of evolution of life on earth, including its cycles of mass extinctions, necessitates a shift from anthropocentrism to "geocentrism" whose scope would be "kin to but more extreme than the Copernican revolution".[52] In *Process Theology as Political Theology* John B. Cobb Jr. famously described the rejection of anthropocentrism by process thought by expressing what seemed on the face of it a standard deep time perspective:

> We are a very special part with peculiar capacities and value. But we came into being at a late point in the evolutionary process and we will some day be gone. That will not be the end of the world, only of humanity.[53]

But in order to claim at the same time that "viewing our species from the point of view of the total natural process does not mean adopting an attitude of calm detachment" what Cobb was clear about was the mistaken (Kantian) premise upon which most anthropocentric views of liberation have been founded. But is the role of religion then simply to be bearers of the new message of a narrative that stretches far beyond the human story, to contextualise human value within this broader story of life itself? Eaton argues that the evolutionary perspective will change attitudes, and thus she argues for a revolution in *aesthetic* as well as ethical stance towards conservation: "When we experience the Earth in its awesome immensities and ingenuity, we are moved to resist its demise."[54]

Yet it would be naive to imagine that deep time thinking alone deepens or radicalises an ethical engagement with climate change. The responses to deep time thinking produce a range of responses, and not always the ones we would wish. For instance, Stephen Clark notes that we must decide whether or not we think our future will be short and tragic or long and glorious. But people also opt for the "presentist" stance – a way of contextualising our fleeting finitude by giving it infinite existential meaning, as in Blake's "infinity in the palm of your hand, eternity in an hour" or even Paul Tillich's "eternal now". And it is a paradox of recent origin that whereas religion once taught us of the brief frailty of our lives in comparison with God's eternity, these days "the religious are eager to believe that the only available Infinite is alongside and in us."[55] When that present experience is tainted by the ubiquitous threat of tipping points, however, what sense can be made of an existentialist deepening of the present? And how could religions deepen their own commitment to the goodness of creation when the present moment, seen politically, is one of wilful blindness and self-deception? Even Clark, who rejects the all too personalistic and commonsensical presentism, expresses like Eaton the value of deep time as the imagination of "immensity" which "awakens in us a recognition of that Infinite which surrounds and confronts us."[56] But contemplating immensity doesn't really help us consider the richness of the challenge described throughout this chapter – one of considering a possible end, and yet affirming our resistance of that end. It would seem that the deep time arguments frequently offered move too smoothly from politics to paleobiology, and the relief with which they do so is palpable. Read, for example, the inimitable pessimism of John Gray:

> It is only in human terms that climate change can be viewed as apocalyptic, however. In the life of the planet, it is normal... Much biodiversity will be lost, but the earth will renew itself. Life will continue and will thrive – whether or not humans are around to see it.[57]

It is indeed a feature of apocalyptic to make crises appear not as natural cycles but the catastrophic irruption of disorder into the otherwise goodness of creation appear outrageous: to summon up images of that which strikes us as unnatural and corrupt, to heighten the discord between what *is* and what is *desired*. The lesson from apocalyptic narrative, nevertheless, is paradoxically to maintain both a sense of deep time (affirming the ultimate "geocentric" truth of the continuing goodness of creation) and resisting its conclusions: of hoping in a better world in which an Edenic balance is somehow restored. Apocalypse would thus seem to echo a sentiment I hinted at in reference to the ambiguous passing of multiple tipping elements: affirming resistance and mourning simultaneously. It would allow the imaginative performance of the cataclysmic end time revelation to purify, transform, and prepare (for impending catastrophe), without taking away any ethical meaning to the present. The intrinsically dualist approach of apocalyptic strategy will of course strike many people as morally abhorrent, dividing the world into forces of dark and light / saved and unsaved. And yet we should remember that what makes the *principle* of apocalyptic imagination a "comic" drama is the goodness of creation itself, and through the agonistic process of crisis, a redemption of that which remains to be good through creation justifies a hope in a new heaven and importantly *a new earth*. For with a doctrine of the *ultimately* good creation the hearers of apocalyptic narrative are encouraged not only to anticipate that goodness but to build it, or to welcome its establishment in their midst.

Conclusions

Let me be clear: it is not my concern to promote an apocalyptic sensibility, in the way that those texts have traditionally inspired political action. Millennialist cultures show no signs of waning and will always represent the temptation to justify violence, bigotry, and catastrophism in the name of imitating Revelation's purifying fury in anticipation of the new millennium. But by associating apocalypticism *only* with this crudely militaristic eschatology we ignore its powerful dramatic description of a state of "in between times" that is very close to our own. That is, an encouragement of our own pathetic engagement with the story. As one of looking *beyond* the terrible tragedy and beyond death; as an inspiration not to let the tragedy simply unfold but to allow it to really transform our agency in the present, here and now, and to resist inevitability. The narrative not only of calamity but of the resurrection of the dead, as well as the living (present in both Daniel and Revelation) introduces a mixing of the apocalyptic and postapocalyptic genres as they are commonly perceived in contemporary culture. For the heightened crisis experienced by the listener is not only of an event of catastrophe, but of the scenario of loss, such as might be experienced by those contemplating the passing of points of no return. And just as we saw Jewish apocalypse as ambiguously eschatological, so the dramatic effects of apocalypse might also be both an ambiguous hope in what salvation might lay ahead, and a need to transform "infinitely" the experience of the present: to make it worth our total commitment.

I am also aware that to overplay the analogy with biblical apocalypse is to stress the latter's reliance on the *agent* of apocalypse – God, broadly speaking – and to pass judgment on the ultimate benevolence of such an agent, and that I have not left space enough to develop this more purely theological question. But by way of suggesting a future opening of such theological – cultural discourse, I can point to a recent suggestion by Celia Deane-Drummond of the uses of a dramatic rendering of apocalyptic for climate discourse. Deane-Drummond suggests that a dramatic engagement with the fear of what persists *through* and *after* apocalyptic crises is not only true to common experience (we do in fact explore this imaginary through all manner of post-apocalyptic film and literary examples). It also harks back to some original theological explorations of the idea that God, through the person of Christ, descends into that experience itself and offers a light of hope within it. Echoing Hans Urs von Balthasar's dramatic depiction of Easter Saturday, in which

> Christ entered into not just Sheol, or the place of the dead, but in a mythical way Christ may be envisaged as entering Hell, and as one who confronted the chaos that is beyond the devastation of creation, epitomised as absolute evil in the world.[58]

Once again, without being able to explore the contemporary resonances with this doctrinal sense of agency (who is Christ, and in what sense might his entering hell be thought of as a liberation for those also caught up in hell and with little hope for its relenting?), we can at least appreciate the pertinence of traditions that are able to imaginatively engage not *only* with a sense of resisting impending danger – points of no return not yet reached, perhaps – but with a redeeming imperative to hope in the transformation of the present that appears *in our eyes* to be already lost.

The illumination afforded by apocalypse for climate change discourse is surely that crisis is all in the telling and enacting. How do we enact a drama that both calls upon the righteous to stay faithful to their selves, and learns to accept a world that is falling away, that embraces the new? The unfolding climate story should not represent a grand meta-narrative either of certain failure (Lovelock's metaphor of Gaia purging herself inexorably of her ill-fated parasite) or of certain salvation through some utopian feat of geo-engineering or other technological hubris. In a very real sense the multiple experiences and perceptions of climate change are marked by continual losses, from the extinction of species, to the loss of individual human lives, to those of languages and cultures. As Anne Primavesi's chapter on Gaian relationality suggests, this experience of loss is one that should reveal our interdependence on the rest of the earth system, not our isolation from it. But it is also marked by attempts at mitigation, reasons for hope and motivations for a striving for a world one is not allowed the guarantee of succeeding. Without this paradoxical tension we are in danger of allowing people to have a burst of frantic activity to save something (in the next 100 months) and then fall into a state of acceptance and mesmeric inevitability – resigned to the kickback of positive feedback mechanisms. Only a discourse as complex and self-contradictory as apocalyptic can promote both – a hope in the world to come as well as an acceptance of certain change including death. Or, to put it another way, to maintain both a raging against the dying of the light, and a mourning of inevitable passing. Both, I would argue, are deeply religious virtues.

Notes

1 Quoted in Jensen, 5.
2 Mcfarlane, 1.
3 Mcfarlane, 1.
4 Buell, 143.
5 McIntosh, 110.
6 McMurray, 1.
7 <http://www.foxnews.com/story/0,2933,244140,00.html>, 2 February 2009.
8 Ricoeur (1984), ix.
9 Ricoeur (1984), ix.
10 Collins (1984), 4.
11 O'Leary, 63.
12 Revelation 13:4.
13 Collins, 84.
14 Revelation 8:6-9:4.
15 Revelation 21:1-2.
16 See Agamben, 62.
17 Berger, xiii.

18 Baudrillard, 119.
19 McMurray, 8.
20 Ricoeur (1985), 23.
21 <http://onehundredmonths.org/>, 12 September 2008.
22 Ricoeur (1985), 24.
23 Matthew 13:35.
24 Revelation 1:1.
25 Derrida (2004), 80.
26 Derrida (1982), 82-83.
27 Derrida (1982), 82-83.
28 Derrida (1982), 89.
29 Bull, 56.
30 Bull, 72.
31 This insight is drawn predominantly from Adela Yarbro Collins, Crisis and Catharsis.
32 O'Leary, 87.
33 Smith.
34 O'Leary, 13.
35 Jensen, 3. Mcfarlane, 1.
36 See, for example, Moser.
37 Lenton, 1.
38 Tickell, 1.
39 Garman, 1.
40 DC Indymedia 1.
41 Quoted in McCarthy, 1.
42 Lynas, 257.
43 For a more in-depth analysis of the ethical questions underlying the wide adoption of the tipping point concept in contemporary culture, see S. Skrimshire, "Points of No Return: Climate Change and the Ethics of Uncertainty", *Environmental Philosophy* 2009.
44 Age of Stupid (2008), directed by Franny Armstrong. <http://www.ageofstupid.net/>, 4 September 2008.
45 Stibbe, 2.
46 Lee.
47 Murray.
48 Schmitt, 3.
49 Murray.
50 Glavin.
51 Murray.
52 Eaton.
53 Cobb, 5.
54 Eaton.
55 Clark, 179.
56 Clark, 193.
57 Gray.
58 Deane-Drummond, 5.

References

Agamben, Giorgio (2005), *The Time That Remains: A Commentary on the Letter to the Romans*, trans. Patricia Daley, California: Stanford University Press.

Attfield, Robin (1999), *The Ethics of the Global Environment,* Edinburgh: Edinburgh University Press.

Baudrillard, Jean (1994) *The Illusion of the End,* California: Stanford University Press.

Berger, James (2000) *After the End: Representations of Post-Apocalypse,* London: University of Minnesota Press.

Buell, Frederick (2004), *From Apocalypse to Way of Life: Environmental Crisis in the American Century*, London: Routledge.

Bull, Malcolm (1999), *Seeing Things Hidden: Apocalypse, Vision and Totality,* London: Verso.

Clark, Stephen R.L. (2002), "Does it Matter", in: Ellis, George F.R. (ed.), *The Far Future Universe: Eschatology from a Cosmic Perspective*, London: Templeton Foundation Press, 187-188.

Cobb Jr., John B. (1972), *Process Theology as Political Theology*, London: Westminster Press, <http://www.religion-online.org/showchapter.asp?title=3035&C=2542>, 16 September 2008.

Cokinos, Christopher (2007), "The Consolations of Extinction", *Orion Magazine* May/June 2007, <http://www.orionmagazine.org/index.php/articles/article/268/>, September 2008.

Collins, Adela Yobra (1984), *Crisis and Catharsis: The Power of the Apocalypse,* Louisville: Westminster John Knox Press.

Collins, John J. (1997), *Seers, Sybils and Sages in Hellenistic-Roman Judaism,* New York: Brill.

— (1984), *The Apocalyptic Imagination: An Introduction to the Jewish Matrix of Christianity*, New York: Crossroad.

DC Indymedia (2008) "Human Extinction within 100 years claims scientist", <http://dc.indymedia.org/newswire/display/114245/index.php>, 16 September 2008.

Deane-Drummond, Celia (2009), "Beyond Humanity's End: An Exploration of a Dramatic versus Narrative Rhetoric and its Ethical Implications", presentation to *Future Ethics 3: A World Without Us? Imagining the End of the Human*, Manchester 16 January 2009.

Derrida, Jacques (1982), "Of an Apocalyptic Tone Recently Adopted in Philosophy", *Semeia* 23, 63-97.

— (2004), "The Spirit of the Revolution", in: Derrida, J. and Roudinesco, E. (eds.), *For What Tomorrow: A Dialogue,* trans. Jeff Fort, Stanford: Stanford University Press.

Eaton, Heather (2008), "The Revolution of Evolution", paper presented to the EcoThee conference, Chania, Crete, May 2008, (*forthcoming in Vol. 3 of Studies in Religion and the Environment*).

Ereaut, G. and Segnit, N. (2006), *Warm Words: How are we telling the climate story and can we tell it better?* London: Institute for Public Policy Research.

Fiddes, Paul (2000), *The Promised End: Eschatology and Theology in Literature*, Oxford: Blackwell.

Garman, Joss (2008), "The truth is, we're fighting for survival", *The Guardian* 8 August 2008, <http://www.guardian.co.uk/commentisfree/2008/aug/08/kingsnorthclimatecamp.climatechange>, 4 September 2008.

Glavin, Terry (2006), *The Sixth Extinction: Journeys Among the Lost and Left Behind,* New York: Thomas Dunne Books.

Gray, John (2006), "Letting Climate Change Happen", *New Statesman* 29 May 2006, <http://www.newstatesman.com/200605290019>, 28 May 2009.

Jensen, Derek (2006), *Endgame, Vol. 1: The Problem of Civilisation,* London: Seven Stories Press.

Jonas, Hans (1984), *The Imperative of Responsibility: In Search of An Ethics for the Technological Age,* London: University of Chicago Press.

Kermode, Frank (2000), *The Sense of an Ending,* Oxford: Oxford University Press.

Lee, Martha (1995), *Earth First! Environmental Apocalypse*, New York: Syracuse University Press.

Lenton, Timothy et al., (2007), "Tipping elements in the Earth's climate system", *Proceedings of the National Academy of Sciences,* November 2007, <http://www.pnas.org/content/105/6/1786.full.pdf> 6 September 2008.

Lynas, Mark (2007), *Six Degrees: Our Future on a Hotter Planet,* London: Fourth Estate.

McCarthy, Michael (2006), "Environment in crisis: 'We are past the point of no return'", *The Independent* 16 January 2006, <http://www.independent.co.uk/environment/environment-in-crisis-we-are-past-the-point-of-no-return-523192.html>, 24 September 2008.

Mcfarlane, Robert (2005), "burning question", *The Guardian* 24 September 2005, <http://www.guardian.co.uk/books/2005/sep/24/featuresreviews.guardianreview29>, 1 September 2008.

McIntosh, Alastair (2008), *Hell and High Water: Climate Change, Hope and the Human Condition,* Edinburgh: Berlinn.

McMurray, Andrew (1996), "The Slow Apocalypse: A Gradualistic Theory of The World's Demise", *Postmodern Culture* Vol. 6, No. 3.

Moser, Susan C. and Dilling, S. (2007), *Creating a Climate for Change: Communicating Climate Change and Facilitating Social Change*, Cambridge: Cambridge University Press.

Murray, Leo (2008), "Wake up, Freak Out, then Get a Grip", <http://wakeupfreakout.org/wakeup.html>, 24 September 2008.

O'Leary, Stephen D. (1994), *Arguing the Apocalypse: A Theory of Millennial Rhetoric,* Oxford: Oxford University Press.

Polkinghorne, John and Welker, Michael (eds.) (2000), *The End of the World and the Ends of God: Science and Theology on Eschatology,* Pennsylvania: Trinity Press International.

Ricoeur, Paul (1984), *Time and Narrative, Vol. 1,* trans. Kathleen McLaughlin and David Pellauer, London: University of Chicago Press.

— (1985), *Time and Narrative, Vol. 2*, trans. Kathleen McLaughlin and David Pellauer, London: University of Chicago Press.

Rowland, Christopher (1982), *The Open Heaven: A study of apocalyptic in Judaism and early Christianity,* London: SPCK.

Schmitt, Gavin (2006), "Runaway Tipping Points of No Return", *Real Climate* 5 July 2006, http://www.realclimate.org/index.php/archives/2006/07/runaway-tipping-points-of-no-return/, 24 September 2008.

Skrimshire, Stefan (2009), "Points of No Return: Climate Change and the Ethics of Uncertainty", *Environmental Philosophy*, Fall edition (forthcoming).

Smith, Oliver (2009), "Living the Pathos of the End", paper presented to the Future Ethics Workshop 3, Manchester, 2009, <http://www.arts.manchester.ac.uk/lti/projects/religionandclimatechange/futureethics/workshop3/workshop3reports/starterpapers/fileuploadmax10mb,150498,en.pdf>, 29 January 2009.

Stibbe, Arran (2009), "Sustainability past the point of no return", paper delivered at *Future Ethics 3: A World Without Us? Imagining the End of the Human* The University of Manchester, 16 January 2009, <http://www.arts.manchester.ac.uk/lti/projects/religionandclimatechange/futureethics/workshop3/workshop3reports/starterpapers/fileuploadmax10mb,150502,en.pdf>, 1 February 2009.

Tickell, Oliver (2008), "On a planet 4C hotter, all we can prepare for is extinction", *The Guardian* 11 August 2008, <http://www.guardian.co.uk/commentisfree/2008/aug/11/climatechange>, 4 September 2008.

Contributors

Sigurd Bergmann holds a doctorate in systematic theology from Lund University and works as Professor in Religious Studies at the Department of Archaeology and Religious Studies at the Norwegian University of Science and Technology in Trondheim. His previous studies have investigated the relationship between the image of God and the view of nature in late antiquity, the methodology of contextual theology, visual arts in the indigenous Arctic and Australia, as well as visual arts, architecture and religion. He is chair of the "European Forum on the Study of Religion and Environment" and ongoing projects investigate the relation of space/place and religion and "religion in dangerous environmental/climatic change". His main publications are *Geist, der Natur befreit* (Mainz 1995, Russian ed. Arkhangelsk 1999, rev. ed. *Creation Set Free*, Grand Rapids 2005), *Geist, der lebendig macht* (Frankfurt/M. 1997), *God in Context* (Aldershot 2003), *Architecture, Aesth/Ethics and Religion* (ed.) (Frankfurt/M, London 2005), *Theology in Built Environments* (ed.) (New Brunswick/London 2009), *In the Beginning is the Icon* (London 2009), *Så främmande det lika* ("So Strange, so Similar", on Sámi visual arts, globalisation and religion, Trondheim 2009) and *Raum und Geist*: *Zur Erdung und Beheimatung der Religion* (Göttingen 2010). Bergmann was a co-project leader of the interdisciplinary programme "Technical Spaces of Mobility" (2003-07) and co-edited recently *The Ethics of Mobilities* (Aldershot 2008), *Spaces of Mobility* (London 2008), *Nature, Space & the Sacred* (Farnham 2009), *Religion, Ecology & Gender* (Berlin 2009), and *Religion and Dangerous Environmental Change* (Berlin 2010).

Hardy Brix is an eminent painter who has travelled for a long time along the sailing routes in (sub)arctic and Antarctic waters. He is educated from the Arts Academy in Århus, Denmark, where he lives. Brix has contributed to decorations in several public institutions. His works have been exhibited at many places in the Nordic countries and they are represented in three major Danish museums. In his own words he circumscribes his method and work, often in water colours: "My method is simple; I work without sketches or other aids, and put on the initial thin layers of paint with a spatula, a bit carelessly – a couple of colour clashes in the middle of many haphazard shots. It has all been set in motion and will gradually become integrated in the painting. I meet gradually more resistance in the painting process, and, as I reach a point of despair, I look at my palette thinking that somewhere in it is the colour that means redemption. The transitions between the elements of nature – the sky, the mountains, the glaciers, the sea, the ice – are all treated as recollections of feelings, of weather, solitude, despair and isolation, all coming together as the painterly code on the surface of the painting." (<http://www.hardybrix.dk>)

Davide Brocchi was a student of Umberto Eco, took an M.A. in social sciences at the University of Bologna and is lecturer at the Institute for Environmental and Sustainability Communication of the Leuphana University in Lüneburg and at "Ecosign – Academy for Ecologic Design" in Cologne. The social and cultural dimension of sustainable development is the gravitation point of his theoretical and practical activities. In this field he has published several articles. Assistant at the European Institute for the Media (Düsseldorf) from 1998 to 2000, he coordinated until 2004 the cultural network of the NGO Attac Germany. 2005 he founded the international platform "Cultura21" which nets social and cultural researchers, journalists, and artists oriented to the guiding idea of sustainability.

Dieter Gerten holds a Ph.D. in freshwater ecology from Potsdam University and is now a senior researcher at the Potsdam Institute for Climate Impact Research (PIK) in Potsdam, Germany. For more than a decade he has done interdisciplinary research in varied water-related fields such as climate change impacts on lakes, global freshwater resources and scarcities, vegetation-water interactions, global carbon cycle, biosphere dynamics, water and food security, and religious aspects of water in present times and in European history. He has published almost 40 peer-reviewed papers in scientific journals and about a dozen books and book chapters, respectively. Together with Sigurd Bergmann he is presently editing a second book on functions of religion in global climate and environmental change. Dieter Gerten is also teaching environmental and hydrological courses at the University of Potsdam, the Humboldt University of Berlin, the University of Basel and the Polytechnic Eberswalde. He is leading a number of national and international research projects, and is an Executive Committee member of the European Forum for the Study of Religion and the Environment (EFSRE), member of the International Association of Hydrological Sciences (IAHS), and member of the European Drought Centre. In 2007 he was also an Expert Reviewer for several reports of the Intergovernmental Panel of Climate Change (IPCC).

Petra Hansson is currently a Ph.D. candidate at the Department of Curriculum Studies, Uppsala University, Sweden. Her thesis is to be regarded as a constituent part of the project *Encounters with Nature and Environmental Moral Learning – a multidisciplinary study of educational practice for sustainable development in the perspective of environmental education, ethics and environmental history* financed by the Swedish Research Council. The overall aim of the project is to investigate students' prerequisites for learning of environmental moral values. In relation to this, Hansson's research interests concern the meaning of art and the reading of literature within education for sustainable development. She has published *Readings for Climate Change – Ecocriticism and Climate Change Education Research*, (forthcoming, Southern African

Journal of Environmental Education), *Förortens färger: Natur och kultur i Johannes Anyurus "Det är bara gudarna som är nya"* (in: S.L. Schulz (ed.), *Ekokritik: naturen i litteraturen,* Uppsala 2007), *Klimatet och litteraturen: Klimatkritisk läsning av Robinson Crusoe* (forthcoming in D.O. Kronlid (ed.), *Klimatdidaktik: Att undervisa för framtiden*, Stockholm).

Thomas Heyd holds a Ph.D. in Philosophy and teaches in Philosophy and Environmental Studies at the University of Victoria, British Columbia, Canada. He has a keen interest in the interaction among values, belief systems and cultural approaches to the biophysical environment. He also has worked on Japanese aesthetics, vision quests and pilgrimage, and on the significance of Indigenous rock art for the relationship between human beings and land. He recently co-chaired the session on "Culture, Values and World Perspectives as Factors in Responding to Climate Change" at the *Climate Change: Global Risks, Challenges and Decisions* Congress, Copenhagen, March 2009 (<http://climatecongress.ku.dk/>). Articles include "Exploring cultural dimensions to climate change" (in: W.N. Adger, et al. (eds.), *Adapting to Climate Change: Thresholds, Values and Governance,* Cambridge 2009); "Rapid natural changes: A perspective from environmental philosophy", (*Quaternary International*, Vol. 173–74, 2007, 161–165); "Nature, Culture and Natural Heritage: Toward a Culture of Nature" (*Environmental Ethics* 2006, 339-354); "The Unity of Bashō's Narrow Road to the Interior: Vision Quest and Pilgrimage", (*Poetica* 2006); "Rock Art Aesthetics: Trace on Rock, Mark of Spirit, Window on Land" (*Journal of Aesthetics and Art Criticism* 57, 1999, 451-58). Books include *Encountering Nature: Toward an Environmental Culture* (Aldershot 2007), *Recognizing the Autonomy of Nature* (Columbia University Press 2005), *Analyse der Bedingungen für die Transformation von Umweltbewusstsein in Umweltschonendes Verhalten* (co-author with Schluchter et al., Bonn 1996), and *Aesthetics and Rock Art* (ed. with John Clegg, Aldershot 2005).

Timothy B. Leduc holds a Ph.D. in Environmental Studies from York University in Toronto, Canada. Questions about the nature of interdisciplinary environmental research and the complementarity between scientific and cultural understandings are central to his past and present research. After being a Social Worker in northern indigenous communities for a number of years, he undertook doctoral research that looked at climate change through a dialogue between Western climate research and Inuit knowledge. This research has been published in the journals *Climatic Change*; *Ethics and the Environment*; *Worldviews: Environment, Culture, Religion*; and *Ecological Economics*. His book entitled *Climate, Culture, Change: Inuit and Western Dialogues with a Warming North* is being published with the University of Ottawa Press in Fall 2010. He is currently conducting SSHRC Post-Doctoral research at the University of Toronto's Centre for Environment that is considering dominant historic trends in the practice of interdisciplinary research and pedagogy in Canada's academic Environmental programs.

Michael Northcott holds a doctorate in theology and sociology from Sunderland University. He was formerly Associate Professor in the Southeast Asia Graduate School and Field Education Director at the Seminari Theologie Malaysia in Kuala Lumpur. He has worked for the last twenty years as Professor of Ethics at the School of Divinity in the University of Edinburgh, UK. His research focuses on the interface between theological ethics and the human and natural sciences. He has published widely on "urban" theology and the theology of work, economic ethics, globalisation, war ethics, modern apocalyptic, ecological ethics and bioethics. He is currently working on the ethics of technology and human work for a book entitled *The Morality of Making* for Oxford University Press. Another long term writing project concerns the theological and ethical implications of humanly caused species extinction. He is also continuing research on the ethics and politics of anthropogenic climate change and the ethics of political economy and is planning a volume on the ethical implications of climate change science. Recent publications include *An Angel Directs the Storm: Apocalyptic Religion and American Empire* (London 2004, 2007), *A Moral Climate: The Ethics of Global Warming* (London and New York 2007), *Theology After Darwin*, (edited with R. J. Berry, Carlisle 2009) and *Diversity and Dominion: Dialogues in Ecology, Ethics, and Theology* (edited with Kyle Vanhoutan, Portland, ON 2010).

David Olsson Kronlid holds a doctorate in theology from Uppsala University and works as senior lecturer at the Department of Curriculum Studies at Uppsala University. His thesis *Ecofeminism and Environmental Ethics* (2003) offers an analysis of ecofeminist ethical theory. He is currently engaged in studying climate change justice and education for sustainable development and is co-editor of Southern African Journal of Environmental Education volume 26 (2010) on climate change education. He has published chapters in *Spaces of Mobility: Essays on the Planning, Ethics, Engineering and Religion of Human Motion* (London 2008), *The Ethics of Mobilities: Rethinking Place, Exclusion, Freedom and Environment* (Aldershot 2008) and *Gendered Mobilities* (Aldershot 2008).

Konrad Ott studied philosophy and history at the University of Frankfurt, and holds a Ph.D. under the supervision of Jürgen Habermas. In the 1990s, he worked as researcher and lecturer at the universities of Tübingen and Zürich, where he specialized in ethical theory, discourse ethics, methods in applied ethics, environmental ethics, and technology assessment. Since 1997, Ott is professor for environmental ethics at the University of Greifswald. This chair is integrated in the interdisciplinary study program "Landscape Ecology and Nature Conservation", where Ott works in close cooperation with ecologists, environmental economists, and experts in international relations. His ongoing fields of research are discourse ethics, environmental ethics, theories of sustainability, justice, and ethical aspects of climate change. From 2000-2008, he has been a member of the German Environmental Advisory Council, which counsels the German Government in environmental affairs. He contributed to three full reports and to five special reports. In the full reports of 2002, 2004, and 2008, Ott was lead author for the scientific basis and the ethical principles of the chapter on climate change and energy policies. He is also a member of the Board for Sustainable Development of the Protestant Church of Germany and full member in the "Deutscher Rat für Landespflege", and has contributed to the *White Paper on the Ethical Dimension of Climate Change* (Donald Brown et al., Pennsylvania State University, Rock Institute). Ott has published several articles and studies on ethical aspects of climate change and related topics. Among his publications are: *Theorie und Praxis starker Nachhaltigkeit* (with R. Döring, 2.ed. Marburg: 2008), *Reasoning Goals of Climate Protection: Specification of art. 2 UNFCCC* (with G. Klepper et alt, Bad Neuenahr-Ahrweiler 2004), and *Klimavorhersage und Klimavorsorge* (with M.Schröder et alt, Berlin 2002).

Anne Primavesi pioneered a course in Environmental Theology and held a Research Fellowship in the subject at Bristol University. She became a Research Fellow at The Centre for the Interdisciplinary Study of Religion, Birkbeck College, University of London. She is currently a Fellow of the Jesus Seminar and Westar Institute, Wllamet University and a Founding Research Fellow of the Lokahi Foundation, London. Her published books include *From Apocalypse to Genesis: Ecology, Feminism and Christianity* (Burns and Oates and Fortress Press 1991); *Sacred Gaia: Holistic Theology and Earth System Science* (Routledge 2000), *Gaia's Gift: Earth, Ourselves and God after Copernicus* (Routledge 2003); *Making God Laugh: Human Arrogance and Ecological Humility* (Polebridge Press, 2004); *Gaia and Climate Change; A Theology of Gift Events* (Routledge 2009). Among numerous articles she has contributed the entry "Theology and Earth System Science" (*The Field Day Anthology of Irish Writing*, Vol 1V, Cork University Press 2002); the entry on "Gaia" (*Encyclopedia of Religion*, 2.ed. New York 2004); the entry on "Ecofeminist Theology" (*Cambridge Dictionary of Christianity* , Cambridge and New York 2006); the entry on "Ecology" (*The Blackwell Companion to the Bible and Culture*, Oxford, 2006); "The Preoriginal Gift – and Our Response to It" (in: *Ecospirit: Religion, Philosophy and the Earth,* eds L. Kearns and C. Keller, New York 2007); "Can Gaia Forgive Us?" (in: *Earthy Realism: The Meaning of Gaia*, ed. M. Midgley, Exeter 2007); "Transforming The Theological Climate in Response to Climate Change: Jesus and the Mystery of Giving" (in: *Nature, Space and the Sacred,* eds S. Bergmann, P.M. Scott, M. Jansdotter Samuelsson, H. Bedford-Strohm, Farnham 2009); the entry on "Gaia" (*Berkshire Encyclopedia of Sustainability, Vol. 1, The Spirit of Sustainability* , eds W. Jenkins, W. Bauman, Great Barrington Mass. 2010).

Stefan Skrimshire is a postdoctoral research associate in philosophy of religion at The University of Manchester, and is politically active in social and environmental movements. He researches and teaches the roots of apocalyptic thought in political life and continental philosophy. He is broadly interested in the persistence of three core ideas in contemporary political cultures: apocalypse, eschatology and utopia. He is the editor of *Future Ethics: Climate Change and Apocalyptic Imagination* (Continuum, 2010); director of a documentary film, *Beyond the Tipping Point?* (2010) and the author of *Politics of Fear, Practices of Hope* (Continuum, 2008). He has led a 3 year research project in Manchester on Religion and Climate Change.

INDEX

PLACES

Persons

SUBJECTS

Studies in Religion and the Environment / Studien zur Religion und Umwelt
published on behalf of the European Forum for the Study of Religion and the Environment by Sigurd Bergmann

Sigurd Bergmann; Yong-Bock Kim (Eds.)
Religion, Ecology & Gender: East-West Perspectives
The understanding of nature is at the heart of European self – understanding, while in Asia the terms of life and energy play a similar central role. Globally, many institutions and movements have made the protection of the environment and climate a top policy priority. Given the urgency of environmental problems the lack of reflections about the human and especially the spiritual dimension of environmental problems is striking. Environmental – and – climatic change transforms not only culture, politics, and economy, but also religion. Religious traditions have on the one hand always been dependent on human ecologies; on the other hand they vibrantly affect our perceptions of nature and sociocultural practices with(in) it. If life and religion change dramatically at present, how could religion make a change? How are religious and ecologic processes gendered, and how can ecofeminism deepen our understanding of justice? What are the life – enhancing spiritual resources in the East and the West? How can Christian theology contribute to the necessary eco – cultural revolution ahead of us? And how can Buddhist, Taoist, Confucian and Christian spiritualities cooperate in a common space and future? Questions like these are reflected upon by scholars of religion and theology from Korea, Canada and Scandinavia. Their chapters emerge from an international workshop, which was arranged and convened by the editors 2007 in Yecheon on the Korean countryside and in Seoul. The book offers the 1st volume in a new series established by the European Forum for the Study of Religion and the Environment.
Bd. 1, 2009, 208 S., 29,90 €, br., ISBN 978-3-8258-1901-9

Sigurd Bergmann; Heather Eaton (Eds.)
Ecological Awareness
Exploring Religion, Ethics and Aesthetics
The past years have seen an ecological development in religions that is staggering. These efforts are responses to difficult local and global ecological problems, with an increased awareness that religions need to be alert, engaged and active partners in the work for a sustainable future. *Ecological Awareness* – with 17 authors from theology, religious studies, biology, sociology and philosophy – explores how religious practitioners have become increasingly aware of ecological challenges. The book considers aspects of ecological awareness: personal, social, political, religious and ecological. It sheds new light on an essential function of belief systems, which function not only as cognitive and moral systems, but emerge from and affect our human body and its mode of perceiving our milieu and ourselves within it. The book contributes to an increasing awareness of our embeddedness in larger life processes, as well as the awareness of life as a gift.
Bd. 3, 2010, 208 S., 29,90 €, br., ISBN 978-3-8258-1950-7

LIT Verlag Berlin – Münster – Wien – Zürich – London
Auslieferung Deutschland / Österreich / Schweiz: siehe Impressumsseite

Nijmegen Studies in Development and Cultural Change

edited by Prof. Dr. Frans Wijsen, Prof. Dr. Ruerd Ruben and Dr. André van der Ven

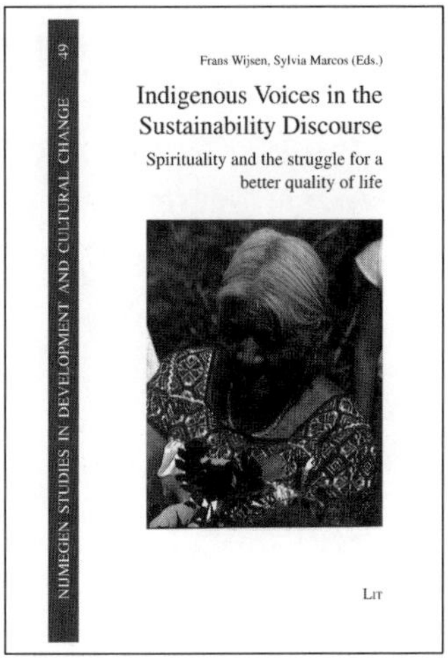

Frans Wijsen; Sylvia Marcos (Eds.)

Indigenous Voices in the Sustainability Discourse

Spirituality and the struggle for a better quality of life

This book is the outcome of a research program conducted by an international and multidisciplinary team of 20 scholars on the relation between indigenous spirituality and sustainable development. It reveals that there is a clash between the developmental views of modern scientific knowledge and traditional indigenous knowledge, each claiming to be better able to contribute to sustainability than the other. Some authors put more trust in modern science and others in indigenous spirituality, yet others occupy a position in between. Whatever their position, all authors hold that in principle evidence-based research can show which knowledge claim is more appropriate to bring about a better quality of life.

Bd. 49, 2010, 440 S., 44,90 €, br.,
ISBN 978-3-643-90053-1

Naomi van der Meer

Believers in the Universal Church

Processes of self-identification among Catholic immigrants of African descent in the Dutch religious landscape

Since the late 1980's the major cities in the Netherlands are home to a growing number of Christian congregations, founded in particular by Africans. This study investigates whether, for African Catholics in the Netherlands, religion contributes to their segregation, integration or assimilation. It contains a comparative case study of the experiences, patterns of affiliation, identity discourse, and social relationships encountered among African Catholics in three faith communities: a Dutch, an international, and an African parish respectively.

Bd. 48, 2010, 312 S., 29,90 €, br.,
ISBN 978-3-643-90051-7

Sanne Derks

Power and Pilgrimage

Dealing with Class, Gender and Ethnic Inequality at a Bolivian Marian Shrine

"Power and Pilgrimage" is an in-depth anthropological study of life at a Bolivian pilgrimage site. It focuses on the experiences of pilgrims and how, in their Marian devotion, they express and learn to live with the various inequalities they experience in everyday life. Issues of poverty and class inequality lead them to approach the Virgin of Urkupiña to support them in their quest for economic betterment. Another social inequality that comes to the fore is based on gender: in particular Bolivian women seek Mary's support to deal with violence and oppression in their homes. Finally, ethnic inequalities are discussed by analysing the dance processions in honour of the Virgin, since these reflect contested ethnic identities.

Bd. 47, 2009, 232 S., 24,90 €, br.,
ISBN 978-3-643-90014-2

LIT Verlag Berlin – Münster – Wien – Zürich – London

Auslieferung Deutschland / Österreich / Schweiz: siehe Impressumsseite

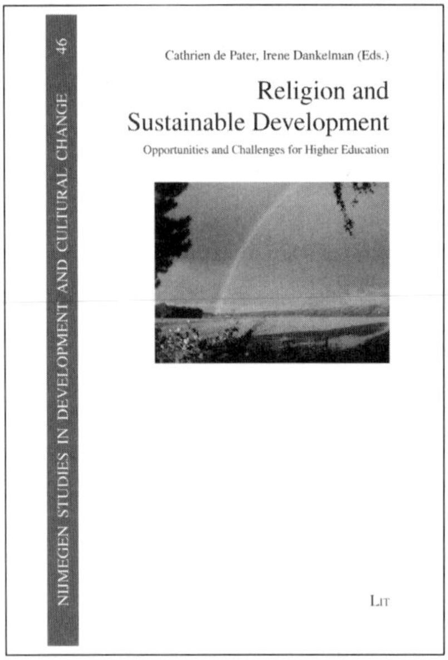

Cathrien de Pater; Irene Dankelmann (Eds.)
Religion and Sustainable Development
Opportunities and Challenges for Higher Education

How are the different worldviews and religions as well as their alleged contributions to sustainable development incorporated in the academic disciplines of religious studies and theology? This theme is explored from the perspectives of Jewish, Islamic, Buddhist, Hindu, modern and indigenous spiritual traditions worldwide. Cases are presented from Western Europe, North America, Malaysia, India, and Africa. This publication shows that the crosscutting of the arenas of sustainable development and religion in higher education constitutes a diverse yet little explored field, and presents the first steps to document its bounty.
Bd. 46, 2009, 184 S., 19,90 €, br.,
ISBN 978-3-643-90017-3

Carl Sterkens; Muhammad Machasin; Frans Wijsen (Eds.)
Religion, Civil Society and Conflict in Indonesia
Bd. 45, 2009, 200 S., 19,90 €, br.,
ISBN 978-3-8258-1261-4

Austria: Forschung und Wissenschaft – Literatur- und Sprachwissenschaft

Alwin Fill; Hermine Penz (Eds.)
Sustaining Language
Essays in Applied Ecolinguistics

Can language diversity be sustained on this earth? Can the resource language be used in a sustainable way? The essays in this volume address language loss in the age of globalization and discuss how language construes our view of Nature and Environment. The volume is also a plea for using an ecolinguistic perspective in school teaching. Kann die Sprachenvielfalt der Erde erhalten bleiben? Kann die Ressource Sprache nachhaltig genutzt werden? Dieser Band thematisiert den Sprachverlust im Zeitalter der Globalisierung und zugleich die Macht der Sprache bei der Konstruktion von Natur und Umwelt. Das Buch ist auch ein Plädoyer für die ökolinguistische Perspektive im Schulunterricht.
Bd. 5, 2007, 312 S., 19,90 €, br., ISBN 978-3-8258-9858-8

LIT Verlag Berlin – Münster – Wien – Zürich – London
Auslieferung Deutschland / Österreich / Schweiz: siehe Impressumsseite